The Lean 3P Advantage

A Practitioner's Guide to
the Production Preparation Process

The Lean 3P Advantage

A Practitioner's Guide to the Production Preparation Process

Allan R. Coletta

CRC Press
Taylor & Francis Group
Boca Raton London New York

CRC Press is an imprint of the
Taylor & Francis Group, an **informa** business

A PRODUCTIVITY PRESS BOOK

CRC Press
Taylor & Francis Group
6000 Broken Sound Parkway NW, Suite 300
Boca Raton, FL 33487-2742

Version Date: 20120208

International Standard Book Number: 978-1-4398-7911-5 (Paperback)

Library of Congress Cataloging-in-Publication Data

Coletta, Allan R.
    The Lean 3P advantage : a practitioner's guide to production, preparation and process / Allan R. Coletta.
        p. cm.
    Includes bibliographical references and index.
    ISBN 978-1-4398-7911-5 (pbk.)
    1. New products. 2. Production management. I. Title.

TS170.C595 2012
658.5--dc23

2012000647

Visit the Taylor & Francis Web site at
http://www.taylorandfrancis.com

and the CRC Press Web site at
http://www.crcpress.com

To Mom Kate and my family—

the real "Kathryn's Finest"

Contents

Preface

Why 3P?

My first recollection of the impact of flawed engineering design came while I was a young engineer working in Operations in the chemical process industry. An engineering team was developing a new railcar off-loading operation for a hazardous material. Railcars would be rolled onto an unloading spur and lines hooked up in order to pump the liquid material into a storage tank. Two operators had to drag a hose coming from a pump about 50 feet away, connect it to the bottom of the railcar, and then start the pump and transfer the material.

The design engineer, a seasoned veteran with 30 years experience, designed the system to complete the full transfer of the railcar in a few hours. To accomplish this he sized the pumps and related piping and also selected a 50-foot long, 6-inch diameter braided stainless steel hose to make the transfer. This diameter hose was necessary to satisfy the pressure drop requirements on the intake and discharge sides of the pump in order to achieve the recognized pumping rates without "starving" the pump.

Unfortunately the engineer never considered that this massive hose weighed hundreds of pounds and was certainly not flexible enough to easily attach while lying on your stomach, suited up in a rubber protective suit, underneath a railcar. It was an impossible task to ask of typical production operators. Ultimately that design was abandoned before the stainless steel hose was ordered. However, the redesign wasted thousands of dollars and the project delivered late.

Over years spent in operations, I witnessed many examples of poor engineering design in plants across the United States and Europe. Many of the problems were eventually discovered, and the impact was seen largely in terms of costs and timing. In most cases the fundamental inefficiencies created in the initial design were embedded so deeply that only incremental improvements could be made after the fact. This reality creates significant loss in profitability over the product life cycle for materials produced in these production operations. Choices made during the development of these products and operations in terms of material selection, process method, machine limitations, and disregard for maintenance requirements result in gross inefficiencies, estimated to be as much as 75% to 80% of the "baked-in" costs that are nearly impossible to "Kaizen out" afterward.

Leaving the chemical process industry and working in the medical diagnostics business confirmed to me that suboptimal designs were not industry specific. "U"-shaped lines were built with the opening of the "U" facing the wrong direction and hindering resupply of raw materials or removal of finished product. Overdesign and the installation of highly complex and difficult to maintain solutions versus simple "fit for purpose" methods were very common and seen in different plants and operations. The greater abundance of profits and money to invest did not necessarily correlate to better designs, just more expensive ones. The additional regulation of the industry also created a level of conservatism and risk aversion that favored established previously validated approaches over improved new approaches, even when the benefits appeared obvious.

What is *your* "burning platform"? Is it simply a desire to continually improve how products and processes are developed? I suspect for many of us the threat of global competition and unyielding cost pressures are forcing us to consider alternate paths at the risk of extinction. One colleague shared a story of his East Coast plant that manufactured blood pressure diagnostic kits. They were facing significant competition, and profit margins were evaporating. In an attempt to salvage the business they developed a next-generation kit, with a greatly reduced cost. However, their great idea was developed using their standard product development process, and because of the time it took to reach the launch phase, they found in the 11th hour that this part of their business had been sold. The new product was never launched by them. A large portion of the plant was shut down and people lost their jobs.

Discussions with colleagues in every type of manufacturing operation in every conceivable industry revealed example after example of ineffective designs that created inefficiencies and issues that would likely never be corrected due to the cost or time required. Oversizing, undersizing, poor flow, inability to lubricate or clean, lighting problems, access problems, lack of consideration for nonroutine functions, and many other issues simply get missed in the haste to get projects done. No one ever thought to "ask the expert"—the operator, maintenance person, or material handler who physically interacts with the product and production process. It was simply the "engineer's job," often working tirelessly in isolation against unrelenting constraints and with no incentive to chance improvement. Innovation was sacrificed on the altar of expediency and risk aversion.

Small companies have some advantages in developing effective designs over their larger counterparts because communication is generally more open and decisions are made faster. However, they often lack the resources and are not willing to spend the time necessary to work through a defined process. Larger companies have the resources but are hampered by functional silos limiting their ability to develop cross-functional value-creating designs. The ability to take a product from inception on a Research and Development (R&D) department's whiteboard through to commercial product, ready to ship to customers, is highly complex. In every case opportunities are missed.

These phenomena are also seen in *Lean* operations. Typically Lean tools are applied individually rather than as part of a complete system. This often results in suboptimization of processes and operations. Value stream mapping is effective in identifying waste and inefficiencies and pointing to the best places to apply resources for improvement. However, value stream mapping is typically applied to *existing* product lines where all of the original design concepts have been implemented in a defined operating area, utilizing defined equipment and technologies, and optimized for a given lot size. Too often the potential business disruption and overall costs and effort required prevent significant improvement projects from being implemented, and only the most modest improvements are done, resulting in incremental improvements.

Total productive maintenance (TPM) is an excellent Lean process that intends to engage all of the stakeholders in maximizing the utilization of an operating asset. Maintenance, Engineering, Operations, Quality, Technical, and Facilities all influence how well an asset performs. During the course of a TPM event, a cross-functional team cleans and inspects every part of a line or operation and in that process tries to create countermeasures to improve reliability and mitigate failure modes. Very often minor design changes are suggested to improve performance. Unfortunately larger, more sweeping redesigns are not possible because the operations do not have the time, money, or space for dramatic step-change improvements.

All of us who have ever wandered around operating plants have our war stories of the suboptimal designs well-intentioned engineers have inflicted upon operations.

So why do well-intentioned, intelligent, and experienced people develop such poor designs? Why does it take so long to develop new products? Why do so many good designs fail to go further to become great ones? The endemic symptoms are everywhere. It is not exclusive to one industry or geographic location. It does not seem to be associated with skill level or experience. It comes from people who have good attitudes and bad attitudes, in big operations and small ones.

I was commiserating about this with a friend and colleague one day, determined to find a way to build more collaborative thinking and innovation into the front end of designs so that what we delivered *exceeded* expectations and better met the needs of the whole organization. I asked if she had ever seen a process for doing that effectively, a process where all elements of a line operation were considered and built into a highly effective holistic design.

She looked at me, smiled knowingly, and said simply, "3P."

And with that, a journey began.

Acknowledgments

True to the spirit of Lean 3P, this book is a testimony to the power of collaboration, rapid learning, and continuous improvement. Many people have given their time, talents, and wisdom freely to make this offering as accurate, practical, and useful as possible.

In 2009 a colleague and I had an opportunity to present on Lean 3P at the annual Association for Manufacturing Excellence (AME) conference. Afterward Michael Sinocchi, Senior Acquisitions Editor for Productivity Press, came up to me and asked me to consider writing a book on the subject, as there was very little published on the topic at the time. Over the months ahead, I continued to apply Lean 3P and develop a deeper understanding of the benefits. Eventually, I started writing and in the process, learning.

In the fall of 2010 I sent a rough draft of the manuscript to Maura May, a friend and former editor for Productivity Press. She provided honest feedback and encouragement and was a great sounding board for the approach taken in this text.

In early 2011 the manuscript was moving along but seemed to need something to pull the key points together. The idea of Kathryn's Finest was developed as a way to bring the concepts and philosophy of 3P together in a story that the reader might relate to. Carl Jarvis, an amazing designer and participant in numerous 3P events, offered to provide the sketches of the fictitious event and along the way convinced me that holding a mock event might be an even more powerful way to convey the concepts. To conduct even a mock 3P event required a lot of help and planning. Friends and colleagues all willingly volunteered their time and effort on a beautiful Saturday in May to help further the cause of Lean 3P.

My sincere thanks to Jim Burns, Alexandra Coletta, Rose Coletta, Matt Delaney, John Douglas, Charlie Fouraker, Jonathan George, Paul Greenwood, Tuan Huynh, Carl Jarvis, Carol Jarvis, Alan Montross,

Bill Morse, Terryn Mozer, Philip Newman-Sanders, Al Risley, Loretta Stiglianno, Mike Szymanski, Earl Vaughn, Nancy Zeller, and Stuart Zeller. Thanks also to Dana Evans, of Evans Photography, for the excellent pictures that so richly augment the text.

As the draft manuscript was coming together I was able to receive input from some amazing subject-matter experts, and friends. The interchange between them became a true version of "catch-ball," as they challenged, disagreed, and affirmed the many concepts covered in the book. The results of those efforts are reflected in the pages ahead. In many ways this is a collaboration of ideas and practical examples that I hope will resonate with you, the reader.

Andy Johnson is a colleague and Operations Manager at Dentsply Caulk (Milford, Delaware), and a fellow 3P practitioner. He added greatly to the practical examples and pitfalls to watch out for. Drew Locher was one of the first to open my eyes to Lean and the power of continuous improvement. He has continued to serve as an informal mentor to me over the years and had great insight into this book. Maria Elena Stopher has worked with me for many years and was the person who first introduced the idea of 3P to me. She has been an amazing facilitator for many of our 3P events over the years. Ken Rolfes helped us through our first 3P event and provided many of the graphics and examples used throughout the book. His insight into the concepts and practical application of 3P is nothing short of profound. Ron Mascitelli has an amazing insight into Lean product development and project management. He painstakingly went through my text line by line, offering insight and developing the depth of the offering at every step. John Shook willingly read the manuscript and offered suggestions for clarification and positive affirmation. As I reflect on the time each of these people spent helping to make this book more complete, I am humbled by their brilliance, experience, and genuine desire to help bring these powerful concepts to light so that each of us will benefit.

I have been privileged to work for a great company with great management at Siemens Healthcare Diagnostics (Glasgow, DE) and with the finest Engineering and Facilities team in the world. They have challenged me and tested me and helped to refine the practical application of Lean 3P. We hope that our mistakes along the way will help you to avoid some of them.

The great people at the Delaware Manufacturing Extension Partnership, especially Lisa Weis, have also been a great asset to my personal education and helped influence this book.

Lastly, I have to recognize and thank my wife Rosemary, and daughter Alexandra, who sacrificed many hours of "family time" to allow me to burrow away in the mornings, evenings, and weekends to complete this work. Thank you. I love you.

Introduction

Life before 3P

History is strewn with stories of engineering designs that have resulted in expensive and sometimes fatal outcomes. Every story is different, but when the final analysis is done, every design problem had an understandable and preventable root cause.

The famous leaning Tower of Pisa, built in several phases over a 177-year period, has become a major tourist attraction due to the precarious manner in which it leans. To our modern eyes, the root cause is obvious—the 185 foot tall tower is anchored by a 10 foot thick foundation, built on sand, rubble, and clay.

In November 1940, the Tacoma Narrows suspension bridge collapsed into the waters of Puget Sound near Tacoma, Washington, due to sympathetic vibrations caused by the wind blowing from just the right direction at just the right amplitude. Video clips of the bridge show it flexing and heaving like a child's jump rope before it broke apart and crashed into the waters below after only a few months of operation. Better structural analysis would have prevented this catastrophe.

Few people will forget the tragedy of the space shuttle Challenger that exploded just after launch, killing seven astronauts and leaving a world in stunned disbelief on January 28, 1986. The cause of the incident was O-rings that failed to seat properly in the frigid atmosphere, allowing hot combustion gases to leak from the solid rocket boosters and burn through the external fuel tank. The O-rings were never tested at those temperatures, and the tragic result changed the course of the U.S. Space Program.

Anyone who visited Boston in the early 1970s would remember the infamous "Plywood Ranch." The John Hancock skyscraper was an architectural triumph, built with over 10,000 windowpanes as exterior walls. Designers quickly learned that those large, 500 pound, 4 feet by 11 feet panes crashed to the sidewalks below whenever the wind gusted to 45 mph. With no immediate solution to the problem, the glass was replaced pane by pane using sheets of plywood. It took years, millions of dollars, and a scale model study in Massachusetts Institute of Technology's (MIT) wind tunnel to determine the cause and eventually correct the problem.

Most design flaws are less dramatic; however, they can be seen everywhere. The pressures of time, money, and stretched human resources are often blamed for these problems, and they certainly all contribute. However, knowing that those pressures will always exist, how can we improve our probabilities of success?

In fairness, perhaps we should consider some great designs. What about the things that are designed and made so well that people flock to buy them? What is it about the proverbial "better mousetrap" that people will rush to your door to buy? How about the iPod or iPhone produced by Apple that has taken the world by storm and completely changed our thinking about how we listen to music or what a mobile phone should do? These devices represent breakthrough engineering on multiple fronts. Apple's innovation of inexpensive downloadable "apps" that provide entertainment and an incredible variety of practical and useful information right at your fingertips no matter where you are is changing how the world works, learns, and plays.

One of the most unique enablers of the success of the iPhone was its input interface using an "intelligent" capacitance-based system that interprets what you want the device to do based on the way you run your fingers across the screen. This breakthrough technology was developed by Wayne Westerman and John Elias, in their small company called FingerWorks, who had a vision of what this technology could do with the right software interpreting it. But it was not until Apple bought FingerWorks in 2005 that the technology they developed became incorporated into Steve Jobs' vision and the marketing, sales, software development, and other functions within Apple, and a great idea was transformed into a world-changer.

Great designs typically start out as good designs. How do we enable talented engineers and surround them with the right people so that they take good designs and develop great ones? Even though we should not expect every design to be a breakthrough, most designs simply stop when they reach "good" and never stretch further for the chance to become great. Wouldn't it make sense to pursue a process that is designed to capture the many voices that interface with a product and process, and take your good ideas and designs and make them much better with numerous break-throughs along the way?

The pursuit of innovation and creativity is often considered the "holy grail" for new product development. Everyone hopes to be the next 3M, where Dr. Spencer Silver accidentally discovered how to make a low-tack, reusable adhesive that eventually became the blockbuster product "Post-it Notes." But accidental discovery cannot be counted on, and even at 3M, the transformation of the invention to a marketable product took a long and highly integrated journey before it arrived on our desktops.

How do we encourage invention and innovation? How can we foster innovation in a structured way that enables individuals to explore brilliant concepts, and then quickly utilize the collaborative genius of people knowl-edgeable of other functions to give it structure and viability? Is it possible to do this quickly and at a low cost? Even more importantly, can this be developed into a company culture, such that brilliant ideas are nurtured and pursued by cross-functional teams as a normal way of working? If this vision could be achieved, the potential impact would be staggering.

Lean 3P, an acronym for the Production Preparation Process, is a system-atic design approach that encourages creative thinking within the framework of a fairly structured process. It helps individuals think openly about new concepts and innovative ideas, and then provides a support team with many different perspectives to quickly develop the concept and then compare it to other ideas using an established protocol.

Lean 3P design embraces and works in conjunction with all of the proven Lean methods. The essence of 3P is to use people effectively to concurrently develop products and production operations that will create customer satis-faction and lower total costs over the life cycle of the product, while ensur-ing consistent quality throughout an efficient supply chain. The concepts were pioneered at Toyota, but 3P is gaining acceptance in many other indus-tries as a means of driving innovation, avoiding costly mistakes, and creating long-term value.

Lean 3P is like the Plan, Do, Check, Act (PDCA) cycle "on steroids." It speeds around that continuous improvement circle over and over during the course of the event, eventually arriving at a plan that will go forward for further development and execution. The concept of "fail fast and fail cheap" captures the spirit of 3P. Event participants work long hours at a high energy level, constantly brainstorming, try-storming through rapid experimentation, and converging onto more and more effective designs for both product and process, and in the utilization of people.

The high-level goals of Lean 3P are big. Research into the approach to 3P goals at various companies shows differences, but in every case they are looking for radical benefits. Ken Rolfes of KDR Associates (San Diego, California) proposes a "4-¼-4" rule be used when comparing 3P to traditional "monument" design. Traditional thinking suggests building one massive plant that can handle the peak load, with full staffing all the time, and at high capital cost. Lean 3P thinking challenges us to utilize fewer resources, energy, materials, space, people, and equipment. It suggests we create four small cells capable of producing one-fourth of the demand, built for a total of one-fourth of the capital cost, and operating with a fraction of the labor, thus creating a four times multiplier on productivity per labor hour. The 4-¼-4 is a good starting point to set the bar high for the team. Other companies expect a 30% or better reduction in the price-to-earnings benefits of the project. In every case the one constant is a high expectation of tangible economic impact when using the 3P process.

Every project and every production operation have three common factors that influence their success: initial capital cost, ongoing cost, and time. Time seems to be the least variable factor, given that most products and new processes are desired sooner than practical and committed to later than they should have been. Time pressure is a significant factor in most projects. 3P results in quicker launches and shorter time to attain standard quality and standard costs for product coming off a new line.

Capital cost versus ongoing costs is a typical trade-off in projects. Higher automation should mean lower labor. More manual operations tend to cost less initially, but ongoing expenses will be higher. The trade-offs go much deeper, however. An automated line might require a more expensive software technician to troubleshoot it versus an operator or traditional mechanic. Manual lines may rely more on inspection by people versus machine intelligence and vision systems. In 3P the expectations and goals are set high and create a type of crisis within the team as they recognize that without very different thinking it will be impossible to be successful. To be clear, the changes proposed could represent additional risk through the addition of new technologies or processes, but they do not have to. Lean 3P often results in much simpler designs that can lower risk to the project and lower the ongoing costs to support and maintain the operation.

The goal of this book is to demonstrate the value of 3P and cover the key elements of the process so that the reader can begin applying it immediately. It focuses on the practical application of 3P so that you can use it to make the design of new products and operations highly effective. You will see how the 3P Production Preparation Process hears the "voice of the customer," engages all of the stakeholders involved in the product or process, and not only helps avoid poor designs but indeed fosters innovation, taking good designs and evolving them to great. Finally, we recognize that every project and design will be unique. We spend some time developing "why" 3P works so that over time you are able to successfully adjust the process to match your own needs and company culture, while pointing out some simple-to-follow methodologies that are proven to work.

Chapter 1

Lean 3P Design Concepts

What Is 3P?

Lean 3P is an integral element in Lean Product Development; however, it is not a holistic Lean Product Development process in and of itself. Depending on how it is applied, it can expand or contract to meet the needs of the organization either receiving input from other Lean Product Development elements or pushing outward into them.

If Lean can be defined as the engagement of people in the act of continually increasing value creation in our processes, then Lean Product Development must focus on that goal as related to the myriad of processes that take place as clever ideas are translated into successful products in the hands of happy customers. The focus of product development is largely about learning and then applying that knowledge to the targeted application. Market analysis, voice-of-the-customer, technology development, and even project planning and management are critical elements of product development and like all processes, they can be improved by applying Lean thinking.

Lean 3P should be inserted early in the Product Development process, as we'll explain later, and will be used to align all of the many vertical functions in our horizontal value stream, in order to optimize the new product concurrently with the operation that will produce it, in consideration of the people who will interact with it. Readers interested in a broader look at the full Lean Product Development process are encouraged to read *Lean Product and Process Development* (Ward 2009), or *Mastering Lean Product Development* (Mascitelli 2011).

The output of the product development process provides the recipe that sets the stage for a profitable value delivery system, which satisfies the needs of the targeted customer. The 3P (Production Preparation Process) process provides a collaborative and concurrent framework to link the product and process design. Most companies utilize new product development processes that treat product design and process design as independent and discrete entities. Integration of the two improves both and reduces the time and costs required to bring the product to market. Lean 3P is event-based team activity that employs a broad range of stakeholders from many different functional areas, including product design, process design, operations, procurement, and representatives from other internal groups. It could also include customers, suppliers, and outside experts.

If the product is in the early conceptual design stage, the 3P event will focus on understanding customer wants and will work to create numerous prototype models depicting different features that satisfy those wants and needs in different ways, using a rapid, iterative process to get feedback and direction. With every prototype, manufacturability is considered. As the product becomes more defined the balance shifts to design of the process, developing many alternatives for each functional step of the process. Changes to the product, including packaging, are considered as manufacturing alternatives and are evaluated before collapsing down to the better alternatives. Physical prototypes are created for the best alternatives, and every feature of the proposed new process is evaluated and compared. The tension between product design and process design continues throughout the event, and both are refined as the design concepts are tested, improved, and retested over and over again.

3P is a Lean design process that provides opportunity to achieve the lowest overall product life-cycle cost because it encourages innovation and collaboration with all of the key stakeholders very early in product development, utilizing all of the tools in the Lean toolkit. This in itself is atypical because Lean tends to be the approach we take to solve problems, not prevent them. Most of our Lean experience involves Kaizen efforts aimed at improving existing processes. In a very real sense, Lean 3P is no different. We are simply applying the same methodology to the design process. Perhaps the real difference is that we are applying the Kaizen discipline methodology to two processes, product design and process design, concurrently. By avoiding problems and pitfalls very early in the product development process, costs can be minimized, quality can be designed in, and many future headaches are avoided.

This is not intended to imply that new product and process development is simple. The fact is that both aspects of design are very complex and fraught with pitfalls. The people part of the equation is a third dimension that is often overlooked or discounted. The people element includes understanding what will really delight a customer. It also considers other groups who will interact with the product or operation. This includes people like the maintenance person who will be charged with keeping the new line going, and the operator, ensuring that he or she will be fully engaged in value-adding work while running the operation, and many more.

Caution: Lean 3P proactively employs the many elements of the Lean toolkit. To be successful the team must be well versed and experienced in Lean application.

Development of the 3P process is attributed to Chichiro Nakao, a former Toyota group manager and the founder of Shingijutsu Company, a Japanese consulting firm. The accepted meaning of 3P is *Production Preparation Process*. As more and more companies have begun to use 3P this definition seems to have become established. 3P allows a group of people to codevelop a product and the operation that will manufacture it in a way that adds value to both aspects—essentially a robust, well-thought-out production, preparation process.

3P is sometimes referred to as *Product, Process, People*, and this definition probably captures best the spirit of 3P. It focuses and attempts to optimize the collective design of the actual product and the production operation that will produce the product, with strong input from and consideration of the people who will interface with it from all the many different functional areas. Figure 1.1 represents the balance between product, process, and people that Lean 3P endeavors to attain.

The *Product* reference defines the necessary product attributes via analysis with deep understanding of the true functions and features that the customer desires, in conjunction with the impact on its overall manufacturability. We tend to think largely of functional benefits when considering a product, but the way it is packaged, displayed, and labeled can influence customer acceptance, and profitability. Developing a product this way could sound limiting and restrictive, but the process of collaboration and incorporation of other viewpoints should only make for a better product *and* a more profitable one.

Process looks at the means of physically moving materials, interactions of people, and producing product in a commercial operation by first

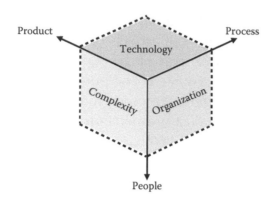

Product   Process
Technology
Complexity   Organization
People

Figure 1.1   The three dimensions of Lean 3P.

developing a full understanding of the actual value-adding features and then considering manufacturing alternatives that best meet those criteria. When Product Development and Operations people achieve a common understanding of the important product features in the context of how that product is made, together they can make intelligent decisions that satisfy both needs without sacrificing functionality or increasing costs. 3P intentionally develops the process with input from hourly workers who will operate and interact with it, gaining a unique and valuable perspective that is normally ignored.

The *People* aspect of 3P considers customer needs, process work flow, accessibility for operator interaction and maintenance, workplace organization (5S), lighting, work heights, changeover processes, as well as material handling, set-up of local material storage (supermarkets) and replenishment systems (Kanban), and a host of other elements that are very easy to consider up front but are often difficult to impossible to incorporate later. The people aspect often has a significant benefit to ongoing costs of operation, but it goes far beyond just operators, mechanics, and material handlers. 3P can utilize all of the stakeholders whose functions touch the product or process, and each one can leave his or her mark on the improved process, baked-in from the beginning. Figure 1.2 shows the possible participants in the Lean 3P process.

Lean 2P

Very often a product has already been defined and is being sold commercially and additional capacity is required because demand is growing.

Sources of Cross-functional Input

Research & Development	Quality Assurance	Process Engineering
Design Engineering	Health, Safety, Environmental	Supply Chain/Logistics
Operations Management	Material Handling	Finance
Technical	Maintenance	Equipment Vendors
Production Operators	Customers	Sales/Marketing
Raw Material Suppliers	Regulatory/Compliance	External Experts

Figure 1.2   Direct and indirect contributors to a typical product value stream.

Product changes then become much more difficult and sometimes impossible because of the risk any change could pose on customer perception or established features. We find ourselves in a position where we need to expand production operations but cannot greatly influence the fit, form, or function of the product. What do we do when limited in this manner and not permitted to touch the product's look, feel, or design? Fortunately Lean 3P design still works, though some optimization will be limited. The ability to use a multidisciplined approach to collaborate on the design of a new process can still yield enormous benefits. When product development and change are very limited, the process is referred to as *2P (Production Preparation)*.

The first exposure I had to 3P was primarily a *Production Preparation* exercise. It was an existing operation that had started out as a development prototype and continued to be used successfully to manufacture commercial product. As the product demand grew over the years new bits of kit were bolted on and added to increase capacity. The control technology employed was becoming increasingly difficult to maintain and we recognized a significant risk of failure. In addition, sales demand continued to grow and was threatening to exceed our production capacity within 2 years. We applied 2P by forming a very diverse, cross-functional team inclusive of Operations, Planning, Procurement, Process Engineering, and other groups, including some of the original senior scientists who developed the product. It was the latter group that proved the most interesting. We will discuss the 3P process in detail later, but suffice it to say that these brilliant people absolutely flourished at the point in the process when people are encouraged to think as they did when they were 12 years old. It was almost as if they had never stopped thinking that way. The other interesting observation was that "tribal knowledge," the combination of biases and beliefs that we passionately adhere to, irrespective of the science or facts behind them, was just as much a part of the senior scientist's mindset as it was the operator who ran the line. During the event, everyone's ideas were challenged. Everyone learned. When it was done a highly collaborative team had bonded together, and together they developed an amazing production operation.

The application of 2P, where a new process is being developed with limited ability to change the product, is actually much more prevalent than 3P. It should be considered whenever additional capacity is needed to satisfy demand or reduce costs. It also applies when a variation of an existing product is added that requires an operational overhaul. In companies new to Lean 3P the leap to concurrent development of both product and process might seem insurmountable. Applying 3P to the development of a new operation may allow introduction of the concepts in a more confined manner, enabling it to eventually expand upstream to the product design processes as the benefits become more apparent to the organization.

Throughout the book we refer to the process as 3P. The truth is that in every event, there is an opportunity to challenge the fit, form, and function of the product and its packaging. With even the most established products, good ideas for improvement will generally be entertained, and considered if they create enough value. Some of the changes might need to be placed on a back-burner to be researched and applied at a later date, but do not assume that product improvements are completely "off the table" even when you are told so. As someone once said, "No is simply the starting point of the negotiation," and any 6-year-old child would agree.

As a rule of thumb, Lean 3P can be used whenever you are designing a new product, making significant changes to existing product features, increasing plant capacity to meet growing customer demand, relocating factory operations, making process improvements to improve product quality rates, or planning capital equipment purchases. As we explore later, the key elements of Lean 3P can also be applied to more modest design changes in a way that adds similar value on a proportionally smaller scale. Figure 1.3 explains when to use Lean 3P.

When to Use Lean 3P
Designing a new product
Increasing plant capacity to meet customer demand
Relocating factory operations
While making process quality improvements
Purchasing capital equipment

Figure 1.3   Events that benefit from Lean 3P.

Why Does 3P Work?

The 3P process sets the stage for engineers to collaborate and share dialogue with a broad cross section of stakeholders before too many decisions have been committed. The term collaboration invokes the idea that stakeholders from the functions somehow gain a revelation and everyone sees the world in the same way. Nothing could be farther from the truth. In Lean 3P the functional representatives are there to ensure that their needs are met. Collaboration develops as the tension between the different functions is resolved after evidence-based debate. During every new product design and new process design we are in a learning mode. We are attempting to gain enough knowledge to move the process in the right direction as it lumbers down the path toward product launch. 3P provides a structure that promotes low-cost experimentation that rapidly increases our understanding of product and process and creates a higher probability of launching products more successfully.

As we consider the creation of value within any product value stream it is quickly recognized that customer value is only created as the product flows horizontally across the value stream. In his book *Gemba Walks* (Womack 2011) Jim Womack points out that while customer value is generated horizontally, businesses are organized vertically, with functional silos that tend to sub-optimize, often at the expense of upstream and downstream functions. Lean 3P seeks to include these functions in a way that allows consideration of functional needs while always maintaining a focus on true customer value.

Separating R from D

What would product development look like if we separated research from development? Research is where invention takes place and where limitations and restrictions are given a back seat to allow brilliant new ideas and concepts to be pursued. Development's role is to take those discoveries, develop a thorough understanding of their attributes, and transform them into a product that creates customer value and delivers profits.

Research and invention is a topic that conjures up images of da Vinci and Edison and other giants of science and technology, and there are many proposed methods aimed at capturing its essence. The debate between individual genius and collaboration continues to wage with no clear winners. When invention is applied with a clear focus on development of new products the field narrows, but there are still many different schools of thought regarding which process renders the best results.

When we think about the creation of a new product, where does the idea actually come from? Who really develops it? Who are the people who influence what a product will achieve both functionally and aesthetically? Is it the customer or a marketing team? Perhaps it is a sales person who thinks he understands his customer's needs or maybe it evolves out of a corporate think tank somewhere. Possibly it is discovered by accident in a research laboratory as happened with Spenser Silver's discovery of the glue for Post-it® Notes. Maybe it takes the form of biomimicry as in the case of George de Mestral's discovery of Velcro, found after closely examining the burdock seed's tiny "hooks" under a microscope.

However a potential solution to a defined need is identified, when it occurs, the concept for a new product is born. After that discovery another flood of questions arise and vast amounts of new information are required. How much will a customer be willing to pay for this product? What features will a customer value? What elements are simply required and which new features would add value that could differentiate it from the competition?

The impact of all those decisions directly correlates to the cost of producing the product over its life cycle, inclusive of initial capital costs, fixed and variable manufacturing costs, quality, logistics, and overall profitability. With more questions than answers and many different approaches available to manufacture the product, how does a company go forward? Who makes the critical decisions, and how do the decisions get handed off to all of the stakeholders? Do the stakeholders have any voice or influence over the many aspects of a product or manufacturing operation that affect them? In most cases that opportunity to influence is limited, yet all of these decisions have a direct bearing on the time to market and ultimate product profitability.

One company shared a story where a new product was developed using a very expensive specialty reagent as one of the key raw materials. After going through U.S. Food and Drug Administration (FDA) approvals and launching it commercially, it was discovered that the cost of the final product was unfavorable, driven by the expensive reagent. It was further discovered that they were using less-purified versions of the same reagent material in other applications successfully at a cost that was one-fortieth of the grade they had selected. It took more than a year to replace the expensive reagent, work through the FDA required revalidations to demonstrate equivalency, and eventually get the new version of the product into the marketplace. Although obvious after the fact, why was it not discovered and tested earlier? Speculation was that the scientist in research and development (R&D) had the expensive reagent on hand and believed it might have some inherent benefits to the product, but never did comparison testing, and never evaluated the cost impact, and nobody else questioned the decision.

Could this happen in your company?

Horizontal Development

Lean 3P takes us through a logical progression that intertwines creativity, innovation, and open thinking. It takes the product development process and stretches it horizontally across all of the functions in the value stream that touch it, synchronizing needs and wants along the way. This concept of sharing ownership throughout the entire product development process is a huge benefit. What does your company practice? If R&D is developing the product as part of their established role, how much influence do your other functions have on their output? Perhaps they hand off to a Product Design group and additional product features are developed. If we agree that many of the ongoing life-cycle costs are built into a product during these early stages, looking at it from only one perspective at a time sequentially has obvious limitations. Figure 1.4 depicts the advantages we see when extensive collaboration between functions starts early.

Rapid Learning

3P makes the design of products and operations tangible by using physical models typically created with sticks, cardboard, and other inexpensive, easy to work with materials. When more detail is required, the use of stereo lithography might be utilized. When possible it is preferred to construct

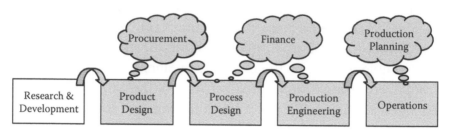

R & D Controls Progression of the Product Development Process

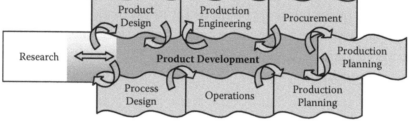

All Stakeholders Own the Product Development Process

Figure 1.4 Vertical product development through silos versus concurrent horizontal product development.

full-scale models; however, when this is not possible or practical, scaled down prototypes are still hugely valuable. The prototype models allow participants to see and touch what the final design will eventually look like. They enable teams to rapidly explore numerous options for every process step and just as quickly collapse those options down to the very best. The prototyping process also enables product design options to be explored in order to evaluate end-product viability, cost, and manufacturability.

3P considers every point of view and allows for a holistic design where the risk of change is embraced by all functional groups, gaining alliance and support to make radical improvements. Virginia Mason Medical Center (Seattle, Washington) presented their use of 3P on a tabletop to design their cancer treatment center at an AME West Region event in 2006. They later used a full-scale model to design the patient treatment delivery room. (Example provided by Ken Rolfes, June 2011.) Figure 1.5 represents a typical approach to new product development.

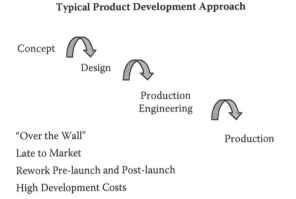

Figure 1.5 Product development suboptimized through functional silos.

All people exhibit preferential bias, a type of intellectual filtration system that validates new concepts and ideas in light of our prior life experience, including our belief systems, and physical, mental, and even spiritual understandings. When acting as an individual contributor, this can skew our thinking and affect our output. Preferential bias tends to amplify those concepts that match our past experience and reject those that differ. Combine these tendencies with the pressure to reduce time-to-market, or move on to the next project, and it is no wonder that designs rapidly focus on point solutions instead of developing alternatives that might prove to be superior. This is one reason that collaboration with others can be so beneficial in helping us expand our thinking and avoid these natural blind spots.

Our natural tendency is to repeat the things that worked and avoid those that did not. However, in real life, feedback mechanisms are often limited or nonexistent. The product or process design groups may never know how well their ideas worked once the product has gone into production. If a new product struggles in the marketplace it can always be attributed to over-inflated sales forecasts, higher operating costs, delayed launch schedules, lower-quality rates or inadequate understanding of customer requirements, and the finger-pointing goes on. Lean 3P engages a cross-functional team throughout the development process, and the product and process design is developed collaboratively from the start and continues all the way through to product launch. This allows ideas to be tested, challenged, and developed at every step, and feedback is supplied from all functional areas in rapid learning cycles.

THE MICROSOFT® KINECT™ PHENOMENON

In November 2010 Microsoft® introduced the Kinect™ for their Xbox 360® gaming system. Within 4 months it became the fastest-selling consumer technology product of all time, with over $10 million in sales. The Kinect was a new game controller that utilized infrared and color cameras to sense the location and movement of users within the space of a room. Microphones captured user commands and blocked out background noise, and most impressively it sold for a mere $150. Previously the type of functionality found in this device would have cost tens of thousands of dollars and would have lacked the integration that Microsoft built into the Kinect. Immediately hackers around the world began working to crack the system and begin applying this revolutionary technology to robotics and many other product types. Microsoft initially resisted, and then eventually embraced the development of their Kinect by the outside technical community. In the July 2011 *Wired*® magazine that reported on this phenomenon, they quoted Hector Martin, a Spanish hacker who was the first to crack the Kinect. "I think it's sad that most companies can't see the value of their products outside their initial idea. There are millions of people who might have better ideas they would never have thought of."

If this is true for one product by one company, why would the same problem of overfocusing not apply to an even greater degree to a product being developed by a single person, or a single department in a given company? Collaboration with customers, suppliers, and all facets of the enterprise can only help to make better products. Lean 3P helps achieve that and enables it to be done rapidly.

A colleague from a large dental supply company cited that one of the most powerful aspects of their 3P events was having all of the stakeholders and decision makers together in the same place for the weeklong event. The progress that was made in a week exceeded what would normally have taken 3 to 6 months to accomplish, and the quality and magnitude of the results were dramatic. The ongoing collaboration of their team continued to reap the benefits of the Lean 3P process long after the product had gone commercial.

Time and cost pressures also force us to minimize innovation and go with tried and true solutions, even if we appreciate that they are suboptimal. Design personnel are encouraged or even forced to narrow the scope of change. Fear of failure and possible retribution is also a driver that fosters the status quo, even when better solutions sometimes seem apparent to all.

As we move through the process steps used in a 3P event, we see why it is so helpful in developing breakthrough solutions and driving innovation, and how it establishes an organizational safety net that permits the participants to take calculated risk in order to gain substantial benefits.

Differences between Traditional Product and Process Design, and Lean 3P Design

There are scores of different approaches available to develop new products and processes. The intent here is not to minimize them or trivialize the benefits obtained using them, but to understand that they have limitations and express how Lean 3P attempts to build upon the best attributes of other development processes to generate extraordinary results. The term *traditional* reflects a broad categorization of many of the commonly utilized product development approaches. The nature of 3P is inherently different than most other design methodologies. Let's compare some attributes of these approaches as they relate to product and process design (Table 1.1).

Due to the complex nature of new product and process development, functional experts are required in every facet of the project. The formal project management processes break up product and process design into logical focus areas, and those are doled out to individual process engineers or design engineers to develop. The vast majority of their work is often done in isolation while sitting at a desk. People draw from their prior experience and new ideas they have been exposed to, evaluating risks versus expediency, while trying to meet the unrelenting project timeline. Most of their effort goes into ensuring that capability and capacity requirements are met. Depending on their personal styles or preferences, others may be consulted, but the engineers have sole responsibility and ownership for developing the concept and design. Normally the design is done using highly intensive computer-aided design (CAD) programming (often using the more imposing three-dimensional CAD modeling that is becoming standard). During this process tolerancing and mating of key parts is commonly done, requiring more time and effort, and it results in a huge investment in what is likely a suboptimal design.

Lean 3P design is different from the traditional design process because it encourages all of the major stakeholders to have ownership in the entire

Table 1.1 A Comparison between Traditional Product and Process Design and Lean 3P Design

Traditional Product and Process Design	Lean 3P Design
Product development is exclusive to the product design function, with a developed product handed off to process design. Concept and design reviews are used to obtain limited input from other functions.	Product design leads product development with *concurrent input* by Process Design and other stakeholders. The manufacturing process is considered in conjunction with product development.
Customer input into new design varies with company-specific application of "voice-of-the-customer" (VOC) comparative methodology.	Customer input is based on VOC, with comparisons of *mock prototypes* and used in an iterative process whenever possible.
An individual process engineer works independently drawing from previous experience and new ideas he or she was personally exposed to. There is some collaboration largely with other process engineers within the functional silo.	A cross-functional team works as a group to explore options and develop innovative approaches based on the collective experience. There is *extensive collaboration* across all functional areas.
Desktop design is the norm with occasional visits to process areas. Detailed drawings are created, often using 3D CAD programs requiring a large resource and investment.	Shop floor design, actively engaging a team of people in developing *physical prototypes* rapidly. Detailed design follows after the total process is optimized.
There is a tendency to reuse existing designs to reduce risk and the time associated with modifying existing drawings versus creating new.	A cross-functional team casts a wide net of *alternative solutions* and then quickly evaluates and narrows them down again and again to develop "best of breed" solutions.
Capacity and capability of the new product and operation are the focus. Peripheral issues and benefits are largely based on individual experience or develop later during concept or design reviews.	Capacity and capability of the new product and operation to meet takt time are key along with flow, safety, people interfaces, ergonomics, quality, buffer and total inventory, and other Lean elements addressing concerns of all stakeholders.

(continued)

Table 1.1   A Comparison between Traditional Product and Process Design and Lean 3P Design (Continued)

Traditional Product and Process Design	*Lean 3P Design*
Formal concept and design reviews are utilized to integrate product with process capability. Due to the extensive effort developing CAD designs, concept and design reviews are often defensive. There is a reluctance to change.	Detailed CAD design comes after prototype models are built and refined through an iterative process. Design and concept reviews take place after the event and are now more of a formality.
Process improvements come *after* process capability is developed. Time is not spent refining the process beyond the ability to perform the essential process steps. Ancillary design elements are often not recognized by the individual or small team completing the work.	Process is developed and *waste is eliminated* at every step via the 3P progression. Very few design improvements should be needed after launch. Considerations for maintainability, quick changeover, and other important efficiency features are built in.
Incremental improvements are made based upon established processes.	*Dynamic change* is fostered, shifting established paradigms.

product and process design, and it allows them to try things. The mantra of "fail fast, fail cheap" for vetting new concepts and approaches was never truer than with 3P. It does not get much cheaper or quicker than rejecting a cardboard prototype developed in a few hours by an energized team.

Time Commitment of the Stakeholders

The level of involvement by the full 3P team will vary depending on the phase of the project and the need for collaboration. The initial 3P event is intensive and creates the foundation for the product and new operation. It mandates complete immersion by the full team. The Lean 3P event is generally a weeklong activity. A 3P product and process development project continues for months longer, completing after the launch and after the new operation is generating product that meets standard quality and costs.

The highest level of 3P has full-time teams for the term of the project. This approach might be required for very extensive projects such as an entire new factory adding new, dedicated support functions. In those cases the 3P team becomes the project team. Most companies who use 3P apply a core of full-time project members and a larger group of part-time project members. The full-time group tends to be mostly engineering personnel, due to the need for developing the design details, drawings, and bid packages. The larger team dedicates a portion of their time to project support. This time commitment could take the form of a daily update meeting or a standard weekly progress meeting with additional review meetings necessary when the need arises. The entire team is responsible for progressing the design, managing project risk, and addressing the inevitable problems as they develop. This occurs using the same methods applied during the 3P event. The team develops new alternatives or revisits earlier options. The alternatives are refined, prototyped, and evaluated to converge on a new best choice. Concurrently updates are made to the schedule and costs.

Do not underestimate the power of the ongoing team participation. In my personal experience with 3P and in every case study I have looked at there is a direct correlation between ongoing team involvement and the overall success of the project. The ongoing cross-functional teamwork allows the whole organization to be aligned with the project throughout the entire development process. It stops the traditional problem of Operations receiving an operation that Engineering tossed over the wall to them, fraught with problems that they now had to fix or live with. Not only do most problems get identified and resolved early in the process, but it builds ownership for the new operation at every step of the way.

Recently I was participating in a tour of our facility with some visitors, standing in one of the new operating areas, when a woman who ran the Planning and Scheduling group quickly walked over to one section of the line and with a beaming smile announced, "I designed this part of the line." When everyone looked at her with puzzled expressions, she stood pointing to a fairly complex part of the machine and continued, "I designed it with my team during the 3P event." And she was right. That is ownership.

Chapter 2

Lean 3P's Place in the Product Development Process

Applying Lean 3P to New Product Development

Voice of the Customer

If the goal of all new products is to create something that a customer is willing to purchase and thus generate a profit, understanding a customer's needs, desires, issues, and concerns is important. Voice of the customer is a process that can be used to do this by capturing a wide range of customer likes and dislikes based on qualitative and quantitative information. This information can be generated through focus groups, interviews, and surveys and is usually very structured, with an emphasis on comparing current products and alternatives in order to get a sense of customer satisfaction. The information gathered is used to develop a hierarchy of needs and wants that become a starting point for product innovation.

Voice of the customer is a very powerful technique that works well in many applications, especially when the new product will be similar to an existing product. However, voice of the customer also has limitations. Breakthrough products do not always have an existing product to compare with, and customers, like all of us, do not know what they do not know, making a comparative process ineffective. Henry Ford once said, "If I had asked people what they wanted, they would have said faster horses." The challenge of breakthrough innovation is its unfamiliarity. A key element in the development of new products involves casting a vision of the possibilities. Lean 3P can help with accomplishing that goal.

Lean 3P can be incorporated very early in the product development process. Developing an understanding of customer needs and wants provides a strong starting point in the product planning process. Even in the early stages of product planning the questions of manufacturability should begin to be asked, initiating the production planning process. As the new product starts to develop functional definition, there may be opportunity to develop physical models of the new product in order to get customer feedback. These physical prototypes will often be nonworking models that simulate the look and feel of the product with minimal development time and costs. Presenting the customer with several mocked-up versions of the new product prototype allows them to look, touch, and feel the possible features and attributes much better than a drawing or written description ever could. The ability to handle it, assess the size and maneuverability of the device, and address the simulated features can happen almost instantaneously when you place it on the table in front of the customer.

If you are making locomotives or jet engines this type of early prototyping might be accomplished using small-scale models. Architectural design firms have used this technique extensively for the same reasons we encourage it now. Most people have a very difficult time understanding and assessing two-dimensional images. With the advent of 3D computer-aided design (CAD) modeling, our ability to understand drawings has increased but still pales in comparison to seeing and touching a real model. It is fascinating to see how effectively very crude prototypes can convey complex design concepts.

Even in the process industries there is potential for using this approach with customers very early in development. Samples created in a laboratory or pilot plant representing a wide range of product attributes are often produced for functional testing by a customer, but they are expensive to make and may not be necessary to get important feedback that will influence the manufacturing process. Does it matter if the product is water white, or is amber acceptable? Should it be crystal clear, or is it okay to be hazy? If the product is a solid would you want it in pellet form or large blocks? Does particle size matter? Would certain side-reaction products have any affect on the application? How pure is pure enough?

In one plant making polyurethane blends I recall expending huge effort and costs processing product through polishing filters, taking out microscopic levels of a reaction by-product, in order to make it crystal clear. Immediately afterward we would turn around and complete it by adding carbon black for the customer application, turning it into a black opaque liquid. In another plant making a powdered polyester material we screened the product to get the particle size into a tight range, generating 20% to 30% rework in the process. Only later did we learn that the range was self-imposed and many customers actually preferred a more random size because it flowed better. How often do we find ourselves going to great lengths to hit self-imposed specifications that our customers do not care about? How many times could a mock sample of another material be used to convey an important product feature to a potential customer, avoiding costs that will often stay with the product over its entire life cycle?

Case Studies

A medical device company was developing a new type of catheter system. Their typical time to develop and launch a new product averaged 30 months for a project of this size. They needed to create both the new product and a new operation to manufacture it. Capital for a new operation was very limited and threatened to kill the project.

The company formed a cross-functional core team represented by Design Engineering, Process Engineering, Regulatory, Marketing, and General Management. They did some reverse engineering of competitive products, acknowledging patented features that could not be duplicated. In short order they cobbled together several nonfunctioning prototypes with different features and met with key customers. The customers handled the models and compared them, suggesting improvements. They made the changes and again sought out the customer's opinions with a few new models to evaluate. With a clear understanding of the "voice of their customer," they used 3P to continue development of the product and the process.

In designing the facility they recognized that a significant portion of the capital cost was tied up in the build costs of a new Clean Room. Utilizing the experience of their team they were able to design a Clean Room at a fraction of the estimated cost. They approached other aspects of the operation the same way, balancing the benefits of automation with the benefits of engaging people in critical process steps.

The resulting product was launched 11 months after the start of development. The capital cost of the new operation was 20% of the amount originally estimated, driven by a creative core team embracing Lean princi-ples of flow, takt, right-sizing, and standard work. The new product was bet-ter than anything the competitors had, and within 18 months the company had captured 25% of the market.

Another company shared an example where their Research and Development (R&D) team had developed a new type of specialized epoxy product that they believed would offer significant customer benefits. They ran a 3P event to further evaluate the product and determine how it might be manufactured. During the weeklong event they had different customers come in to provide input. After several days they realized that the custom-ers had little value for the "improved" features and that the new product was nearly impossible to make on a commercial scale. A unanimous deci-sion was made to send the new product concept back to the drawing board. Initially many people felt like they had failed. However, they had saved months of development time and possibly avoided launching a product that would fail in the marketplace. Instead they had the chance to start again with a better understanding of both customer wants and manufacturing capabilities. They also had the confidence that they would get it closer the next time. Figure 2.1 depicts iterative product innovation using 3P compared to a sequential design progression.

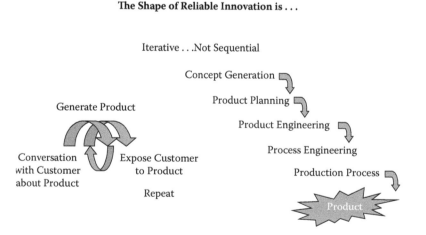

Figure 2.1   Product innovation with 3P versus sequential design. (Adapted from Rolfes, K., KDR Associates presentation, AME Conference, Australia, August 15, 2007.)

Lean 3P within Existing Product Development Processes

Lean 3P is a powerful enabler for invention and innovation because it creates a structure and a process for people to create both independently and collaboratively. However, 3P is not presented as a "one size fits all" means of creating brilliant new products that takes us from "blue sky" to product launch. It might work like that in some instances where a new product is a variation of an established product or in organizations where the same team is inventing, developing, and working together to launch a new product. With additional experience the role of 3P in the full product development will likely expand. For companies new to Lean 3P, the question might be how 3P will integrate into existing product development processes.

Figure 2.2, taken from *Value Stream Mapping for Lean Development: A How-To Guide for Streamlining Time to Market* by D. Locher (2008), demonstrates the flow of the Lean 3P process in development of a new product and process.

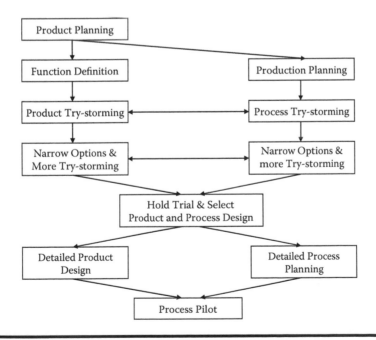

Figure 2.2   The flow of the Lean 3P process. (Locher, D. 2008. *Value Stream Mapping for Lean Development: A How-To Guide for Streamlining Time to Market*, CRC Press, Boca Raton, FL. With permission.)

Phase-Gate Product Development

Many companies employ some type of formal gating process for product development. There are numerous versions of this methodology available in active use, and they all attempt to systematically move a new product from conception to launch by applying structured project management techniques. Credit is given to the National Aeronautics and Space Administration (NASA) as one of the first to apply a phased review process to its development efforts. The Advanced Product Quality Planning (APQP) process was developed by the "Big Three" U.S. automotive makers in the 1980s and it utilizes five phases. The Project Management Institute and others have phase-gates with different attributes and benefits. All attempt to apply structure to the product development process.

Companies sometimes have too many new product possibilities, and with finite resources only the very best ideas can be driven through to launch. The phase-gate process seeks to selectively pass the products with

DESIGN FOR MANUFACTURABILITY AND DESIGN FOR ASSEMBLY

Design for Manufacturability (DFM) and Design for Assembly (DFA) are considered by many to be precursors to 3P. Recognizing that the lifecycle costs of a product are largely established during the early product design stage, a formal review process was established between Product Design groups and Process Design Engineering in order to develop products that were more easily manufactured at a lower cost. Creating a formal design review process lowered product costs and reduced much of the iterative rework between the functions, avoiding delays to launch. DFM focuses on material costs, processing steps, and design tolerances.

Design for Assembly is similar to DFM with a focus on handling, reducing parts, and assembly time and complexity. Both methods were developed and improved by numerous organizations during the 1960s and 1970s. In the 1980s Geoffrey Boothroyd and Peter Dewhurst are credited with developing the methodologies further, using computerized versions of the techniques and expanding the application of it into the automotive, electronics, and numerous other industries. Boothroyd and Dewhurst developed a combined process named Design for Manufacturability and Assembly (DFMA) that is still utilized.

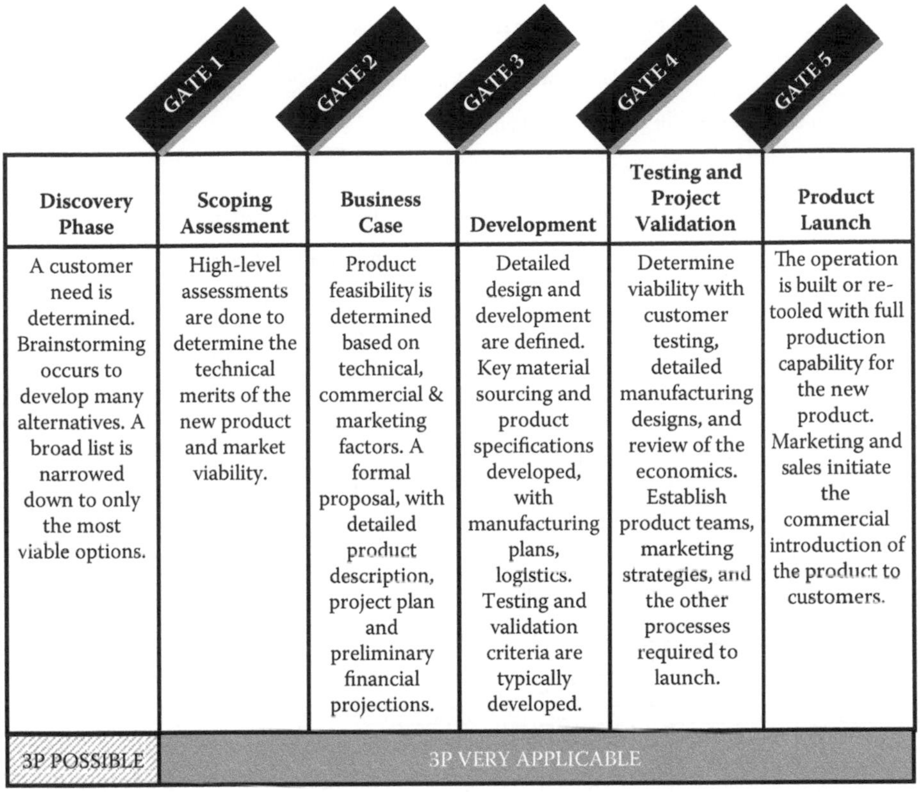

Discovery Phase	Scoping Assessment	Business Case	Development	Testing and Project Validation	Product Launch
A customer need is determined. Brainstorming occurs to develop many alternatives. A broad list is narrowed down to only the most viable options.	High-level assessments are done to determine the technical merits of the new product and market viability.	Product feasibility is determined based on technical, commercial & marketing factors. A formal proposal, with detailed product description, project plan and preliminary financial projections.	Detailed design and development are defined. Key material sourcing and product specifications developed, with manufacturing plans, logistics. Testing and validation criteria are typically developed.	Determine viability with customer testing, detailed manufacturing designs, and review of the economics. Establish product teams, marketing strategies, and the other processes required to launch.	The operation is built or re-tooled with full production capability for the new product. Marketing and sales initiate the commercial introduction of the product to customers.
3P POSSIBLE	3P VERY APPLICABLE				

Figure 2.3   Integrating Lean 3P into a phase-gate product development process.

the highest probability of success by culling other lesser product ideas as early in the process as possible. This is accomplished by the inclusion of formal review and decision gates, where a manager or steering team reviews the information, weighs the risks and the business case, and makes a decision. Some new products pass to the next phase or stage where additional resources will be applied and more functions will get involved to progress the product toward the next gate and eventual launch. Lesser product ideas will be killed entirely, placed on hold to allow stronger candidates to progress, or recycled back for more optimization. One phase-gate process involves a six-phase/five-gate system with the sequence as shown in Figure 2.3.

Reviewing the phase gates above, there is potential for incorporating elements of Lean 3P in the Discovery phase. Depending how you use this phase it might be very applicable. If the nature of this phase is very "blue sky" and the purpose is largely to cast a very wide net of new product ideas

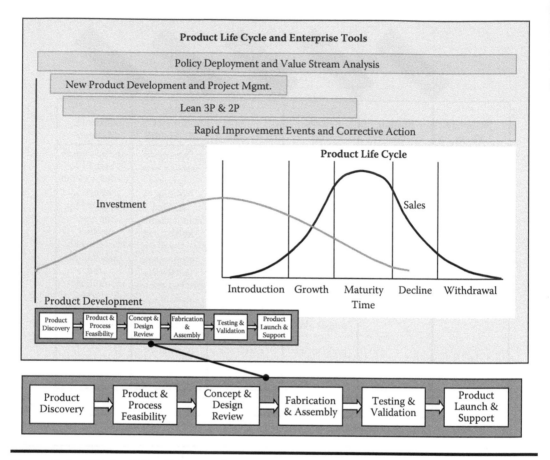

Figure 2.4 Product development investment over the product life cycle. (Adapted from a graphic provided by Rolfes, K., KDR Associates Inc. presentation AME Conference, Australia, August 15, 2007.)

and then narrow them down as you progress toward the first gate, manufacturability might only be considered at the highest level. Once a potential new product reaches the Scoping phase, the cross-functional benefits of 3P should apply and will continue through the Launch phase.

Figure 2.4 represents the place where Lean 3P belongs relative to a typical product life cycle. It is not intended to be all-encompassing with every enterprise tool available, as its focus is on new product introduction from concept through to end of life. The investment we make in developing new products starts out being fairly modest and continues to grow with additional research, prototyping, increased involvement by more functional groups, build-up of manufacturing capabilities, training, and eventually marketing costs and distribution. Most of the investment comes before the first sale and generation of any revenue. Our ability to shorten and reduce the magnitude of that investment period has benefits over the entire product life cycle, and in many cases an earlier time to market will increase market share.

Point-Based Engineering

Though the nature of phase gates is intended to start wide and continue to focus, the practical reality is that many projects progress in a very linear fashion. This is referred to as "point-based" engineering. A solution is proposed and after some development is considered a decision is made to go with that design. The point-solution is progressed until a problem occurs and it is determined to not be "right," and we move on to a new different point and repeat the process. Unfortunately this forces significant changes very late in the development process, when it has the greatest impact on project timing and costs.

Figure 2.5 depicts the difference graphically. Forcing consideration of many different alternatives at the earliest point in time increases the probability of success and minimizes the impact on project costs and timeline.

In his book, *Mastering Lean Product Development: A Practical, Event-Driven Process for Maximizing Speed, Profits and Quality* (2011), Ron Mascitelli develops the idea of dramatically reducing risk at the earliest moments in a project. His utilization of activity "burn-down" charts, where

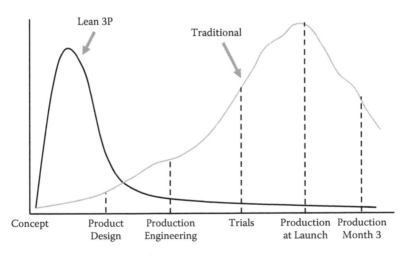

Figure 2.5   Lean 3P optimizes design early versus sequential design. (Rolfes, K., KDR Associates Inc. Presentation AME Conference, Australia, August 15, 2007.)

the highest-risk elements of a project are selectively addressed first, captures the essence of 3P. Mascitelli postulates correctly that if the elements of greatest risk to the project can be mitigated as early as possible, the probability of success increases. Once the major risk is removed, the execution of the project is reduced to the tactical execution of the remaining low-risk steps.

Set-Based Concurrent Engineering

Set-Based Concurrent Design

The last alternative design approach we mention is set-based concurrent engineering (SBCE). This is a method developed by Toyota that involves using "sets" of possible design solutions versus a single preferred option as seen in point design. The different designs within the set compete concurrently, with significant development of each design concept in parallel. The competing designs are evaluated with input from Manufacturing Engineering who influence selection by pointing out designs that are best suited for manufacture. The goal is to find an optimum solution at the intersection of a set of feasible alternatives. When the knowledge gaps are closed, Design Engineering makes selections and the next level of detail is developed in a similar manner. As product design decisions are made, Manufacturing Engineering concurrently progresses the operational design to the level allowed by the characteristics of the complete set currently being evaluated, providing input at every phase based on manufacturing capabilities and preferences.

The selection portion of the process is described as "rapid learning cycles," and Mascitelli (2011) refers to the combined set-based concurrent design and rapid learning cycles as "knowledge-based development." It allows a higher level of refinement in the early stages, without jeopardizing time-to-market because at each phase it allows both product selection and operations design to progress as far as possible within the constraints of the set of options being evaluated. Front-loading the evaluation of many alternatives identifies and mitigates the highest risks at the start of the development process.

Lean 3P embraces some of the SBCE techniques. The development of a wide range of alternatives and the collaboration between Product Design and Process Design are foundational to 3P. However, SBCE takes the development of the alternatives within the set much farther than we do when evaluating 3P alternatives. Lean 3P fosters a broader range of collaboration and incorporation of all of the elements of Lean throughout the development process. SBCE is a progressive process. 3P is a fast-paced, event-based process. Although they bear many similarities, and both avoid the limitations of point-based engineering, 3P may provide benefits to a broader range of product types in less time.

Over the following pages we will begin to develop the 3P process, capturing the many benefits it can bring to us in a way that provides you the insight and confidence to take your next steps.

Chapter 3

Applying Lean 3P Design and Gaining Leverage

Introduction

As soon as a viable product concept is identified and deemed worthy of further pursuit toward a goal of going commercial, a Lean 3P event should be initiated. The reasons for doing this are significant. Improving time to market, meeting product cost and performance targets, and minimizing capital expenditure lead the list.

The concept of gaining leverage has great appeal. The principle of increasing knowledge in the very beginning logically should create more leverage than gaining knowledge after the fact. The former takes advantage of increased flexibility and the relative ease in making change. The latter reflects learning "the hard way," when the cost of change is much more difficult and more expensive. In addition, the scope of what can be changed may be limited. The Lean 3P principles work at whatever point they are applied. Increasing collaboration and developing knowledge by using prototypes to rapidly test a proposed design's effectiveness is good when tackling a problem on an existing operation. They work even better when designing entirely new processes using 2P (Production Preparation). The most leverage comes from using the techniques when concurrently developing a product and process together.

Concurrent Product and Process Design Using 3P

All of the key stakeholders should ideally be involved at some level in all of the stages of the product development process. Process Design and Operations' role in the development of the new product might be limited initially, but as product designs start to become more tangible they should be contributing heavily regarding the manufacturability of the product being pursued. Very early on in development it is about product viability. Will someone want this product? Is there already established competition? How big is the potential market and how much of it can be taken? Does it fill a need better than other existing products or is it truly new and novel? How much is it worth to a customer? What will the customer be willing to pay for it? What are the quality and performance expectations of the product? Can we produce it at a cost that will generate acceptable margins of profit? How quickly can we get it to market? What is the benefit for getting it to market early? What is the penalty for delivering late? These questions and many more must be answered in order to determine if a new product is viable.

For most new products the three critical factors determining feasibility are the *price* customers will pay for it, the *cost* for us to make it, and the *time* to market. Price and cost are usually directly coupled to volume, and there is a balancing act that takes place with every new product when estimating how much more could be sold if the price was lower. Product volume increases tend to drive up plant capital costs, but when realized, lower unit costs. Recognizing that future sales projections are always going to be incorrect to some degree, the risk of overbuilding is fairly high. The third factor is the time required to bring the new product to market. In a competitive world there is always someone else out there trying to do the same things we are—develop better products, take market share, and manufacture at a lower cost. There are many examples demonstrating the benefits of getting to market first and realizing greater market penetration and market share. This holds true even when there are other competing products out there if the perceived value of the new offering is better. When you think of Intel's dominance in the microchip market or Apple's introduction of the iPad you see clear examples of this.

Product features are another factor that drives costs up and slows time to market. If we accept the premise that every product is developed to solve a problem, then we should also recognize that additional features might only have incremental added value to a customer. Yet chasing those features that go beyond "solving the problem" can be a slippery slope that consumes time and adds costs. A new car can solve the problem of enabling transportation. Making it fuel-efficient solves the problem of lowering ongoing operational costs. Designing it to be stylish may be important to a majority of customers, but not all, and we are starting to move away from the primary objectives. A great sound system, satellite radio, built-in Global Positioning System (GPS), and chrome plating on the interior dashboard are all nice features, but they do not hold universal customer value. Many people would be just as content without them, but developing them and building them into the car certainly adds significant cost that affects overall profitability.

Many added features are important to differentiate the product from competition, but at some point they can become incremental, with no benefit in terms of increased sales or customer value. The 3P differentiator is the ability to quickly develop physical prototypes for many types of products. Alternatives can be made and presented to customers for hands-on evaluation and nearly instant feedback. If needed, second-generation prototypes can be created to get even closer to the mark.

Soon after product viability is determined, Process Design, Operations, and Technical must determine if the product can be made at a commercial level consistently, with high quality, safety, and meeting all environmental requirements. They must consider materials of construction, personnel exposure concerns, waste streams, availability and handling of raw materials, changeover issues, space requirements, and many other high-level concerns that establish initial cost estimates and enable a decision to be made to fully understand the business case for the product.

Is One 3P Event Enough?

Depending on the magnitude of the product development scope, the 3P process can be performed over the course of multiple events. Mascitelli develops this approach in his book, *Mastering Lean Product Development: A Practical, Event-Driven Process for Maximizing Speed, Profits and Quality* (2011). He proposes a *3P Design* event that focuses on product design. His second event is a *3P Process* event that develops the higher-level manufacturing plan. The third event is the *3P Production* event, which builds out the detailed manufacturing plan. This approach has merit for many applications.

With highly complex products there may be an additional need to split out key components to do separate 3P events. Consider how an automobile is made. We tend to think of an assembly line, but all of the many components have to be manufactured before they can be assembled. Complex equipment that is really made up of many other complex pieces of equipment, like an automobile, could utilize separate 3P events for the component parts.

I encourage the reader not to get too concerned by how the 3P process could be split up, but to understand the methodology and *why* it works. With that background, you have the understanding you need to determine the best approach to your specific application.

If we think about the product development process as a knowledge-gathering process, start with some questions. Do we know enough about the new product to allow more focus on the process, with a smaller focus on product attributes? Do we have customer input? Is the target market defined and quantified? Do we understand the materials and specifications? How many features have been determined to be "must do's" versus "should and could do's"? Assessing the extent of known product information should lead to a decision of where best to start. If the knowledge and understanding is there but has been developed functionally instead of collaboratively it might be wise to start with a product-focused 3P event to further develop it with input from the other functional groups. The investment in time to first develop the product at this juncture will at worst be redundant, and at minimum will build a solid foundation of knowledge for all of the participants to build upon.

The Role of Operations

A number of years ago a manager that I worked for asked me, "why are we employed here?" My initial response was "to make product" and at that time in my manufacturing career the products were formulated polyurethane foam systems that went into everything from shoe soles to automotive seating. The products were technological marvels combining sophisticated base components with a variety of initiators, catalysts, blowing agents, colorants, and numerous other materials to create a specific type of final product when our customers processed them in their factories. After an uncomfortable pause, he looked at me with a knowing response and said "No, we're here to make money for this company and our products are a means to do that."

I have thought of that many times over the years and have come to embrace a modified version of his mantra. Manufacturing's purpose is to produce consistent, high-quality products that create value for our customers for which they are willing to pay a price that exceeds the profit expectations of our stakeholders.

If this is the optimum role for operations, what should be the role of design? Should it not be creating the recipe that makes this possible? Just as it is impossible for manufacturing to do it alone, it is also impossible for them to do it independently. Using the analogy of a sports team we know that each position must be well defined and played skillfully, but it is the interaction of the players working as a team that wins games.

This philosophical difference in focus calls for a higher level of commitment than just running production and making the numbers. The concept of "creating value" by producing high-quality products that customers are delighted with and willing to pay a premium for takes manufacturing to a new and higher level. But how much *can* we influence? In most plants and in most businesses the things we in Operations can influence are relegated to that which is within the four walls of our factories, with perhaps an occasional venture out into the supply chain.

Most Operations people speak a different language than design people. We do not naturally collaborate well together. Lean 3P provides a framework for that collaboration and the learning that accompanies it. Previously, design capability often resided with a handful of people trained in computer-aided design. With 3P design, anyone who knows how to use a saw or a hot glue gun can convey powerful ideas using simple prototypes, and if they want to improve them it is fast and inexpensive. Everyone can participate.

I include this short section because I believe the people who work in Operations understand the impact of the decisions that get made upstream of them, better than anyone else. We understand the shortcomings and spend a great part of our lives trying to improve them. For many companies the biggest champions of 3P are the Operations people. The very idea of Lean is still lumped together in terms of *Lean Manufacturing*. Operations tend to be the first to reach out to the Lean toolkit to embrace the continuous improvement processes. And we have seen the fruit of that effort.

Applying 3P for the first time to a plant expansion requires a lot of organizational work to educate and build the value case, but Operations people know the leverage it will generate. After the organization gains experience and comfort with the Lean 2P Production Preparation process, it might be easier to drive it further upstream into the early development phase of new products. We know we will see even greater leverage generated if we can collaborate at the start. It is just intuitive—not necessarily obvious, but intuitive. As the number of practitioners expands and as the numbers of case studies get shared, this need for Operations to push from the middle should begin to diminish. Until that time, Operations and Process Engineering will have to continue making the case for increasing our leverage using 3P.

A Leverage Case Study

Ken Rolfes of KDR Associates (San Diego, California) offered an excellent example of the leverage that can be attained using Lean 3P in a company he worked with:

> One 3P project realized a 500% productivity gain by changing the product and manufacturing process designs. The previous paradigm for expanding capacity was to use the "cookie-cutter" approach and simply duplicate existing process designs rationalizing "that was the way we had always done it." Over time, the market pricing had made this approach noncompetitive.
>
> The strength of competition had accelerated a rapid market price decline, resulting in eroding margins and forcing some companies to exit the business. Survival was only possible with a significantly lower product cost.
>
> The cross-functional 3P Team in the plant radically redefined the nature of the needed product design and manufacturing processes. The old process and product designs were replaced with creative, low-cost, reliable, right-sized processes designed to operate on the lean principles of takt time, flow, multitasked operators and quick changeover.
>
> The company knew they had done well with this project, but had the rare opportunity to confirm our success when their competitor acquired the company.
>
> As the Vice President of Operations was taking the acquiring manager through the plant, their conversation went like this:
>
> The Acquirer said as they walked to the front door, "You had better be glad we bought your company."
>
> "And why is that?" the VP asked.
>
> He responded by describing the equipment they were building over the last 2 years at a cost of $2.5 million, that will bring the production cost per unit down to $0.035 per unit when it becomes operational.

The VP responded, "You had better be glad you bought us. I will show you two production lines currently operating at $0.035 per unit and each line cost a quarter million dollars."

At that time the first of the two lines had been operating for 2 years. Even years later, the large machine owned by the acquiring company had not lived up to promise. The complexity of trying to do everything on one large universal machine increased the machine cost, created long changeovers and required heavy maintenance. This approach also increases production risks, with a single operation carrying so much of the available capacity.

Plan your equipment in smaller right sized units and as production requirements increase replicate as needed. Effective application of the 3P methodology can create tremendous market leverage (K. Rolfes e-mail July 2011).

Gaining Leverage

The "Best" Time to Make Improvements

The very best time to make improvements that will affect product quality and profitability is during the earliest stages of development. Applying 3P to the overall design of product and the process gives us complete control of product features, manufacturability, and the role of the people who will interface with it. We already spent some time exploring how and why this works so effectively. It is especially effective when actual customers or suppliers are allowed to participate in this process. Too often we make assumptions on what we think a customer will want or what a supplier will do for us that are incorrect.

A colleague shared an example of a Lean 3P event his company conducted where they asked their key customers to participate, along with their marketing, operations, quality, and the product and process design teams. Together they began differentiating the "needs" and "wants" for the new product they were developing and assigned an estimated dollar impact to each feature based on manufacturing or material impact. This short investment in time correlated to a much simpler process and a product that completely matched their customer's expectations at the right price-point.

Full application of Lean 3P allows for the greatest flexibility in the way the product will look, work, be packaged, and ship to customers. This holistic approach, when done well, can optimize the total customer benefit. Many of the numerous trade-off questions can be asked and challenged before they are imposed on the next functional group. Can we use standard parts or materials? Will it be packaged as a single unit or in a multipack? Does the customer care if the packaging is clear, or what colors are used on the product? How flexible does the line need to be? How will we address a product ramp-up and build a production operation that is neither too big nor too small? Can we build an operation where each processing step is able to meet the same customer demand takt time? Will the operation ramp up easily to accommodate new business? Can it be quickly changed over to make a sister product?

The ultimate design potential comes when the full 3P approach is applied. Using 3P we codevelop a product along with the process that will manufacture it, and involve the people who will operate and interface with it. 3P is the gold standard for development, and it creates an unmatched leverage factor that yields returns over the life cycle of the product.

The "Better" Time to Make Improvements

When the opportunity arises to build a new operation the potential exists to make important improvements. This often happens when new capacity is required to meet the growing demand of an existing product. It can also occur in organizations that maintain the practice of having a separate product development process where Research and Development or Design Engineering creates the new product independently and then hands it off to Process Engineering to design the new operation. Although some opportunity is missed, there is still great value in applying a collaborative team approach to design of a new operation. As noted earlier, this is sometimes called 2P (Production Preparation). Applying 2P provides opportunity to make operational improvements *before* the line is built and the space committed and the people are brought in to run it. This approach assumes that you already have an established product that cannot be dramatically changed and only the Process and People elements can be influenced.

When used to expand capacity for an established product, 2P events allow us to reflect back with experience on the current operation and make dramatic improvements. Those step-change improvements *always* lead to benefits that are impossible to do after the operation is built and making commercial product. Once the new operation is running, the ability to shut it down to make improvements becomes more difficult. The added cost to de-bottleneck the process requires a lot of justification.

In another example, a supplier of silicon semiconductor wafers was contacted during a Lean 2P event that was being conducted to build a new operation for an existing product. Previously the vendor had supplied the wafers in eight-up plates that were processed and eventually broken into individual wafers for the product application. During the event it became obvious that handling individual wafers had significant advantages, but some of the original developers of the product stated that the vendor would either not be capable of providing individual wafers or the cost increase to do it would be prohibitive. The team was ready to abandon the opportunity when the Procurement team member stepped in and said that he had just spoken to the vendor and learned that they could easily do it, and there would be no additional cost. In fact, all of their other customers received their wafers that way.

Lean 2P, while not nearly as effective as 3P, still has enormous impact on quality, costs, and the role of the people who will interface with the new operation, which is nearly impossible to attain after the operation is built and operating. The leverage created by applying resources here should yield benefits many times greater than making improvements later to the completed operation. The ability to make better decisions based on rapid learning in a collaborative process creates opportunity as depicted in Figure 3.1.

The "Necessary" Time to Make Improvements

Most of the time Operations and Process Engineering are limited to making modest improvements or upgrades to existing equipment and processes. We apply Lean thinking and Lean tools to make improvements, but typically the ability to redesign and make improvements at this level will be incremental with respect to affecting operational costs, which translate directly to standard product costs. Once the product is established it becomes much more difficult to change. It will be difficult to reduce features or to simplify the product in any way. Customers might perceive a downgrade of features as a devaluing of the product, even if they do not need it, and expect a price reduction, negating any benefit from the non-value-added simplification. In the regulated industries even a modest change to the product or its labeling would require reporting and submission to the regulatory body.

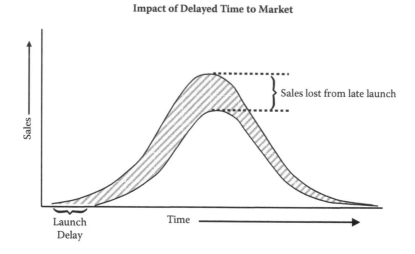

Figure 3.1   **The benefits of shorter time to market.**

The cost of the existing operation is already spent and any improvements will only add more expense. Many operations involve multiple processing steps, so making significant upgrades or de-bottlenecking the least efficient module is useful but tends to just shift the rate or quality-limiting step to the next module. This translates to very hard work with minimal gain.

There is always something more to go after in order to continuously improve. Yield rates are never 100%, and equipment downtime due to failures and very short "micro-stops" can take a heavy toll on performance. Lean methods such as value stream mapping, total productive maintenance, standard work, and quick changeover will provide very real benefits. In my experience, however, the product and process designers had not planned on these losses as part of the ongoing standard cost base when developing the product and production operation, so in reality we were operating below expectations and these important improvements are just getting us back to even.

Summary of Applying Lean 3P Design

Lean 3P design fosters collaborative effort over the entire product development process straight through to product launch and beyond. During the 3P event we focus an intense collaborative effort over a defined period of time, employing rapid learning prototyping. 3P navigates all of the participants through a process that sets us up for success based on a common understanding. Risk is evaluated holistically from many perspectives, and willingness to accept some risk for the potential of significant gain is fostered because everyone has part of it. How many organizations would refuse to try 3P if they understood that a collaborative process could yield exponential benefits over the life cycle of the product they are developing? Sometimes we all just need to step back and consider the possibilities.

Chapter 4

Getting Started

The Case for Change and a Necessary Prerequisite

The biggest obstacle to getting started with 3P might be convincing your organization that it is worth the time, cost, and risk to give it a try. Change requires leadership and conviction, especially without a crisis to create a "burning platform." Most people can quickly grasp the concepts and recognize that 3P could have dramatic benefits. However, it can be challenging to overcome organizational inertia. Making a compelling case for change is critical. Consider an appropriate pilot to start with to demonstrate the power of 3P without the risk of a high-profile new product launch. A manufacturing expansion for an established product may be considered a safe starting point. Fortunately the number of case studies and success stories being published continues to increase, and hopefully the barriers to applying 3P will continue to drop.

3P incorporates many of the traditional elements of Lean including Takt, Flow, and Pull; along with the support tools such as total productive maintenance (TPM), 5S, quick changeover, Poka Yoke, and Jidoka. At the heart of the 3P process is the Shewhart Cycle (also attributed later to Deming); Plan, Do, Check, Act (PDCA). During an event the participants continuously apply PDCA, going around and around to quickly improve design concepts. Organizations that are new to Lean continuous improvement concepts should pursue a deeper experiential understanding of Lean before attempting the Lean 3P design process.

All team members participating in a 3P event should be experienced in Lean to be successful. Although the potential benefits will be tempting, Lean

3P design incorporates all of the Lean methodologies, and it will be very difficult to make progress in an event if the participants do not have a firm grasp of what they are, how they work, and when to apply them. A solid foundation in Lean is a necessary prerequisite.

The Flow of a Lean 3P Event

Lean 3P is an event-driven cross-functional team process. Designing a product concurrently with the manufacturing process involves hands-on experimentation using physical models that emulate the product features or equipment. This rapid prototyping process is called "try-storming," and it involves creation of trial designs to see how well they solve product and process challenges. Try-storming, like the name implies, is a hands-on extension of the brainstorming process. Try-storming is a type of prototyping that takes ideas and quickly mocks them up, so they can be evaluated physically. Figure 4.1 represents how Lean 3P uses try-storming in rapid learning cycles to enable knowledge-based convergence of design ideas.

Information

The first focus is on gathering information, to fully understand the problem that is being addressed through this design effort. The information will come from many sources. Information is collected, organized, and made visual and physical where possible, so that all of the participants fully understand and can add to the body of knowledge being developed. The information developed includes the product and process attributes, incorporating previous lessons learned, forecasts and goals, and boundaries and constraints.

Innovation

The second focus is on innovation and exploring the ideas and potential approaches to each product attribute or functional process step, by developing multiple alternatives and approaches, casting a very wide net. Participants develop possible solutions from nature and from industry. An evaluation process is used to converge a wide range of possible solutions to the better alternatives. Innovation is encouraged individually and collaboratively. Rapid learning cycles are employed to explore and develop knowledge throughout this phase of the event.

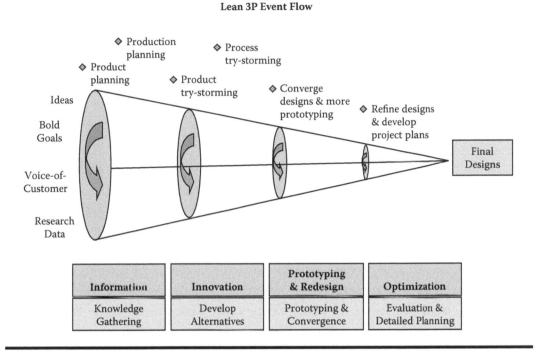

Figure 4.1   The four phases of a Lean 3P design event.

Prototyping and Rapid Redesign

The third area of focus involves developing physical prototypes, applying the better alternatives that the team of participants had selected. More detailed information is developed as the models are built. The prototypes are constructed inexpensively and quickly. Adjustments are made throughout the prototyping process based on visual and tactile observations from the broad spectrum of participants. Eventually the developed alternatives are evaluated and for each functional process step the best alternative from the different prototypes is chosen. From several mock-ups the best attributes are selected, and from them a single prototype is constructed and refined.

Optimization

The fourth focus is on development of the final prototype, building in the detail that will define performance, quality, and costs. The product attributes are harmonized with the process and the people who will operate and interface with it. Project plans and financial estimates are developed and a detailed evaluation is conducted to determine the effectiveness of the proposed product and process relative to the established goals.

Key Participants in a 3P Event

A typical Lean 3P event takes a full week or longer. There are five main groups of people involved directly with the 3P event outcome: Process Owners, Process Experts, Judges, Process Outsiders, and Facilitators.

Process Owners

Process Owners are the leaders responsible for developing the product and/ or the operation that will make the product. They are responsible for the product and the operation required to make it and must agree to the direction the team recommends at the conclusion of the 3P event. The Process Owners should always be participants in the 3P event unless they have complete confidence in the process experts and agree ahead of time to support the group decisions. For companies new to 3P, the process is so dramatic and the potential outcomes so revolutionary, they must participate fully in the event. Some companies refer to this position as Chief Engineeer or Lead Engineer; the person responsible and with the authority to make decisions.

Process Experts

The majority of participants are *Process Experts*. They come from all of the stakeholder organizations that interact with either the product or the operation that will produce the product, and typically 12 to 24 people serve in this role. Larger projects could include more people, and more modest projects could have less depending on the complexity. However, having the right people representing the key functional areas is *always* required. Cross-functional collaboration is the essence of Lean 3P. The methodology captures ideas from people with very different perspectives, exploring, converging, and incorporating them in the final designs.

Process Experts often include participants from the following functional areas:

■ R&D (Research and Development)—understand the technology and the product
■ Technical—understand the complexities of the technologies being employed
■ Supply Chain—understand the impact on distribution and logistics for suppliers and customers

- Marketing and Sales—understand the customers and the market conditions
- Design Engineering—understand the details of the total engineering system
- Process Engineering—understand and can optimize engineering designs
- Production Engineering—understand and can optimize manufacturing processes
- Operations Leadership—understand manufacturability, people impact, and cost structures
- Operators—understand the details that enhance operability, ergonomic issues, and error-proofing
- Shipping and Receiving—understand impact of material delivery, shipments, storage, and the transactions required to track them
- Procurement—understand vendor contacts, confidentiality issues, strategic alliances
- Material Handling—understand the details of the physical movement of materials
- Quality—understand material testing and release, impact of variation, release criteria
- Finance—understand the costing system, capital investment process, impacts on standard costs
- Maintenance—understand requirements for repair and building in reliability
- Skilled tradesmen—sometimes referred to as "moonshiners" in the context of Lean 3P; craftsmen who have a great aptitude for innovation and quickly creating useful prototypes and providing practical design input

The intensity of a 3P event requires a lot of energy and creativity. It is not the kind of training-experience Kaizen that you use to develop people who might be neutral or negative to Lean. People who are selected should be picked because of their optimism, creativity, team orientation, excellent work ethic, and how they react to and value ideas by others. Strong individual contributors with strong technical biases or experience may be needed, but if allowed to become too dominant in the event, they can pull down the level of the entire team resulting in suboptimal results.

Judges

One of the most misunderstood roles in a 3P event is that of *Judges* (*Critical Evaluators*). Part of the 3P process involves bringing in very senior business and technical subject matter experts to review concepts. This will typically involve three to five people and could include the Plant Manager or Vice President (VP) of Operations, a Senior Director or VP of R&D, Site Finance Controller, VP of Marketing or Sales, Director or VP of Procurement or Supply Chain, or others as appropriate to the company. These are the people with a higher-level understanding of the business and customer needs, who often influence or approve resource allocations and capital expenditures. They also tend to be the key decision makers, and their support of the process and team direction is the enabler that allows the breakthrough ideas to gain footing and progress.

Note that the title of this role is not universal and seems to often cause confusion. The term implies a heavy-handed, final decision type of role that passes judgment on the ideas and work of the others. This is not the intent of the 3P Judge's role. Whether you choose to use the term or not, you will find value in having the function of the Judges fulfilled.

The role of the Judges is to bring to the 3P event what Dr. W. Edwards Deming would refer to as "profound knowledge" (W. E. Deming, *Out of the Crisis*, 1982) They are often very senior people who have significant experience in the technologies, the products, customers, the market, or the business. Ideally they are good listeners and teachers who provide a level of insight that captures a strategic view of the issues.

Judges are needed to add balance and perspective to the concepts being presented by the participants and teams. They guide the process in conjunction with the teams who are developing the concepts. They always have input that shapes the direction of the event because they normally have a more strategic view than the others involved. Many Judges enjoy spending more time than requested in the events because of the energy level and creativity. This is wonderful as long as they do not start to stifle the teams. The Facilitators will need to watch for this. If the expectation and role is made clear to the Judges, it will help to mitigate this from occurring.

Judges tend to serve as the "paradigm breakers." Every operation and every product design has certain aspects cemented in. These are ideas and concepts that have always been and always will be no matter how good or bad they are. These are the unspoken boundaries that are never written but are always ingrained in our thinking. "We always do it that way." "The vendor won't change it or if they do, it will cost a lot more." "Our machines only work with this type of material." "The Labor Union won't let us make that kind of change." The list of unspoken rules goes on and on. 3P brings these to light and allows us to raise them up for review, discussion, and resolution.

Judges are a key part of those paradigm shift discussions. More often than not they will initially legitimize the status quo, possibly citing reasons why it has to be the way it is. However, if the team does their homework to investigate and understand the real impact of and the original reason for the current state, most of these paradigms will fall by the wayside. Without the Judges' endorsement to pursue the new approach, most teams would not have the courage to take the risk of addressing the change. The status quo would continue to restrict progress with artificial boundaries.

This endorsement of a proposed change by the Judges enables the biggest gains to be made in every area of the operation. This does not just apply to overriding paradigms. Without the Judges, the 3P event participants will relegate themselves to the safer changes, but with the Judge's support the world of potential possibilities dramatically expands. Can you relocate members of different departments? Should you put separate operations together? Could you make the product on two smaller lines versus one large one and gain flexibility for less cost? Many questions like this will come up and be explored during an event. Having unbiased Judges assisting at those key junctures is invaluable.

Process Outsiders

Most events benefit by having *Process Outsiders* involved, who may come from other parts of the company or outside of the company. People from sister-plants can offer a broad range of information and ideas, potentially having solved similar problems in very different ways. Vendors of key materials, parts suppliers, equipment representatives, and even customers add huge value. Vendors representing key technologies should be requested to be on standby or attend in case a need arrives that they can come in for a short session to demonstrate a new technology. Use some caution with vendors, however, as they can be predisposed to a given solution, especially if it could influence a future sale.

The list of *Process Outsiders* needs thoughtful consideration based upon where potential change is likely to be made. Lining up internal people tends to be straightforward, but lining up external "stand-by" participants can be more challenging. Essential raw material suppliers, key vendor representatives, personnel from sister-plant sites, and any number of others can be very important to keep the 3P event flowing based on the detailed technical knowledge and experience they have. 3P team members sometimes go to supplier locations to see their processes and discuss options for receiving materials in a different form or with alternate specifications. Equipment vendors often visit a 3P event and provide demonstrations of new technologies that one of the prototype teams is exploring. Proactively lining up as many of these as possible, in case they are needed, is very helpful. Note that you may need to have someone from your company's legal staff available to prepare nondisclosure or confidentiality agreements if you intend to discuss your project with people outside of the company. Typically this will not be a problem, but prior planning can save a lot of time during an event.

Facilitators

Excellent and experienced *Facilitators* are another key to success in Lean 3P events. They should be very experienced in the entire Lean toolkit and should have facilitated other 3P events previously. These subject matter experts are helpful in convincing the organization to support you in using 3P initially. For larger groups it is important to utilize two or more facilitators in order to respond quickly to questions and issues that will inevitably arise during the fast-paced event. The essence of a Lean 3P design event is very different than typical concept reviews or design reviews, and good facilitators can keep the event moving, assuage fears and concerns, and ensure that everyone stays engaged in the process.

Some companies utilize very experienced Sensei as facilitators for significant projects. They stay with the full-time teams for the duration of the project and ensure that the 3P process is followed faithfully, with a high level of discipline. The Sensei authoritatively ensure that the problems are brought back to the team to be reprocessed using the 3P criteria, keeping focus on costs and schedule and forcing the team to work the process. A great danger to successful 3P is the tendency for teams to revert back to old ways of designing once the event is over. Good facilitators and experienced project managers, whose only role is to ensure that the process is utilized at every turn eliminate the ability of the team to revert back. This practice is most critical for companies new to 3P, though in reality few companies recognize it until too late. This failure is not fatal and many of the benefits of 3P will be captured; however, it can result in unnecessary rework, schedule slippage, and cost overruns. The 3P process centers on the creative ability of a team to overdeliver on performance at a fraction of the cost. Project risks and problems will happen on every project. 3P led by an excellent facilitator or Sensei will enable the process to work and deliver the targets.

Ahead of the 3P event, facilitators can also refresh the team participating in all of the major Lean processes so that the team goes into the event with a common baseline of understanding. For companies that have a solid background in Lean, a few hours of refresher training might be enough, while substantially more may be needed if the concepts are new. This refresher training should take place several weeks before the actual event to allow people the time to absorb it and accommodate it within their schedules.

Space Required for a 3P Event

Lean 3P is a physical process that requires adequate space. Developing mock-ups of different products with different attributes and features will have very different space needs depending on the nature of the product. Product prototypes, like process prototypes, can be scaled down, requiring less space. The approach used to develop the mock-up models might affect the type of space required. Does your product lend itself to developing prototypes out of wood and cardboard, or plaster or clay? Do you have stereolithography modeling available to you? Consider the type of building that you will be doing when considering the type and amount of space needed for development of product alternatives.

As the focus shifts to development of the process, your 3P event will ideally have enough space to do at least three complete mock-ups of the intended production line or operation. Why is so much space required? Why wouldn't a CAD drawing work just as well? In theory, they could, but most people have difficulty fully comprehending what they see in a drawing versus what it is like in real life. There is something very powerful about seeing and touching something tangible that brings clarity beyond anything else. On some 3P events we actually had operators demonstrate process flow and walk through mock changeovers. Safety representatives can evaluate potential ergonomic issues. Maintenance people can test access to equipment, and material handlers can practice how material is stored and how it will move in and out of the operation. Even the best three-dimensional (3D) computer-aided design (CAD) models cannot compare to the simple creations of sticks and cardboard that characterize most 3P prototypes.

For large-scale operations as seen in many industries, including chemicals, large-scale industrial equipment, petrochemicals, or pulp and paper as examples, scaling might be necessary due to the amount of space that would be required to do full-size mock-ups. In these very large operations the time factor required to create full-size prototypes initially could be prohibitive and impractical. It is best to get as close to real life as possible to enable the participants to fully grasp the layout and relative size. 3P is a highly creative process, however, and space limitations may be a good hurdle that the team should sort through early in the process. Never rule out the possibility of a team coming up with a totally different approach that enables a miniature version of what will become a large process to generate great results.

A case study presented by Dan McDonnell at the 2010 Association for Manufacturing Excellence (AME) conference described a wind turbine gearbox factory that GE built using a rigorous 3P process. The massive gearboxes and the factory size required to build them made use of a table-top model more appropriate for their initial 3P event. He described how they gained significant value by looking at layout and types of equipment in the tiny model, making corrections and improvements before taking it to the next level—a one-sixth scale model of the entire factory built in a large open space. Further evaluations and refinements were done in all aspects of the factory model by the large dedicated team. When they had gotten as much as possible from the one-sixth-scale model, they created full-scale models of the key modules. Space limitations did not allow them to create a complete model of the entire factory at full size, but the GE team came up with a creative solution that captured all of the benefits within the constraints they had to work with.

Every space for a 3P event should include a meeting area with whiteboards, visual projectors, tables and chairs, flip charts, and good connectivity to telephone and Internet services. Like all things Lean, 3P should be very visual. The ability to quickly understand and evaluate new ideas is an important consideration. You will also want space for the three full mock-ups of the operation as previously mentioned, if possible. Within each of the three prototype spaces you need to tape off the allowable space and any physical boundaries ahead of time. Having the prototypes in the same general area is preferred as it allows for easier comparison when moving from one to another to judge the designs. It also makes the development of the final prototype that will be a combination of the best parts from the three much more expedient.

Materials and Tools

Having the right materials and tools available for a 3P event will be a great enabler for getting the most out of the process. The cost of the materials should be minimal, and it will amaze you to see how creative the teams will be simulating complex processes with basic materials. Depending on the type of operation you will be simulating, different types of materials may be necessary, and a good facilitator will be able to help you develop a comprehensive list of materials for your prototypes. The events we ran have benefited most from the following basic materials. Most of them are available at your local home center.

Note that you should estimate liberally when it comes to materials and tools. Leftover materials can always be returned to the store. Teams building prototypes are moving very fast and should not be held up waiting for supplies or because they had to borrow tools from another team. 3P is a high-energy, fast-paced process, with healthy competition generated throughout the process. Be prepared with the right materials and tools ahead of time.

3P MATERIALS LIST OF COMMON ITEMS

- Board stock (good for adding structure or making brackets)
  - 1" × 2" boards
  - 1" × 3" boards
  - 2" × 4" boards (some, but heavy to work with—avoid unless needed)
  - 4' × 8' foam board (used for structure or enclosing or creating simulated parts)
  - 4' × 8' sheets of Luan or other inexpensive material
  - Cardboard—almost any size can be used, but large sizes are best
- Fasteners
  - Nails (different sizes)
  - Drywall screws (different sizes)
  - Staples
  - Duct tape
  - Painter's tape
- Miscellaneous Items
  - Markers
  - 3/4" or up to 1½" PVC pipe
  - "Boneyard" materials—equipment used in your industry that will not likely be used again but not ready to throw out

 – Canvas
 – Pieces of wire
 – Tables
■ Tools and Safety Equipment
 – Safety glasses
 – Gloves
 – Safety shoes
 – Miter saws
 – Handsaws
 – Hacksaws
 – Circular saw
 – Razor knives
 – Staple guns
 – Hammers
 – Drills
 – Hole bits
 – Extension cords
 – Screw guns
 – Hot melt glue and glue gun
■ Children's Books (depicting natural processes, how things work, etc.)

3P Event Charter

Before the event you must develop a *3P Event Charter.* The charter sets the stage for what will be accomplished during the event. It is very similar to the charters developed for other types of Lean Kaizen events with a few elements that are specific to 3P. The charter is a one-page document that summarizes "what is it?", "when is it?", "where is it?", "who will participate?", "what functions will be represented?", "why do we want to do it?", and "what targets do we need to achieve?" Figure 4.2 represents an example of an Event Charter.

The charter should be succinct and kept to a single page. Companies familiar with the A3 methodology, where the "story" is told on a single sheet of A3 size paper, will understand the value of making the key information obvious and to the point. Event Charters are often done on an A3:

PROCESS NAME	START DATE	END DATE
The No-Diesel Diesel Engine Project	May 5, 2014	May 9, 2014
PROCESS BOUNDARIES	**START TIME**	**END TIME**
New product to utilize alternate hydrocarbons in a diesel-style engine	7:30 AM	4:30 PM
	EVENT MEETING LOCATION	
Factory and shop only	South Factory	
Not receiving, warehousing of distribution.	**PROCESS OWNERS**	
	VP Operations: Ron M VP Marketing: Hal M	
WHY DO WE NEED THIS EVENT?	**PROCESS EXPERTS**	
Existing product lines plateau Cost of oil is skyrocketing Hybrids and electric causing market erosion at 5% per year	Product Mgr: John W Design Engr: Kay M Regulatory Mgr: Stan G Director HSE: Jim E Director Operations: Alex K Supply Chain: Paula E Production Super: Joyce W Technical Mgr: Kim G Procurement: George M Quality Super: Narda M Finance: Pam M Operators: Megan B and Alexandra C Facilities: Glenn S	
EVENT TARGETS	**EVENT JUDGES**	
Final product cost under $1,500 USD at rate of 30,000 units per year	Chief Engr: Ron M Research: Maria S COO: Ken R Dir. Lean Sigma: Drew L	
Capital cost limited to $1.2 MM USD	**PROCESS OUTSIDERS**	
Time to market launch 14 months or less	Acme Machine Tool Co: Andrew J	
	FACILITATORS	
	The Sensei Co.: Kit E	
	MEP: Kim K	

Figure 4.2   Example of a Lean 3P event charter.

Process Name—This is the first step and is straightforward. Use a description that is broad enough to enable creative development but useful for identifying the product or manufacturing operation to be created or modified. Choose the naming terminology with some care as it will likely stay with the product and process throughout the life of the project.

Dates for the 3P Event—Set the schedule so that participants can see when they must be there. Be specific with the days and times. Will there be working lunches provided? Set the expectations high and hold people accountable to them. Expect the 3P event to be very intense and require long hours to maximize the effort. Like other Kaizen ("change for the better") improvement events, 3P requires complete dedication and commitment from the participants. Note that the Judges' schedule will be very different as they will only be invited in at select times, and these times can flex due to the ebb and flow of the 3P project pace. Publish the anticipated Judges' schedule under separate cover.

Start Time and End Time for Event—These times are also stated to set expectations. Often longer days are spelled out in the charter so that participants understand and can make accommodations if required. Working lunches are typical and keep people physically in place, while fostering teamwork and allowing the focus to be maintained on the process. Most people look forward to a catered lunch. Sometimes the hours expand during the event due to the amount of work and the high energy level of the teams. Though the days are demanding both mentally and physically, highly energized and competitive teams often choose to work longer hours to continue making progress.

Team Meeting Location—The location for the meeting is identified. Ideally the place for the event will have a meeting-style area plus a larger room to construct full-scale prototypes. If possible it will have a place for snack breaks and lunches. Most events benefit by having food and drinks brought in for the attendees.

Process Boundaries—These need to be articulated very definitively. The 3P event participants need to know how far upstream and downstream they are responsible for developing. Does their responsibility include packaging and logistics upgrades? Are they to design new utility supplies such as steam boilers or electrical substations? Be specific as to what is in and just as importantly what is *out* of scope. This section can also include *boundary conditions*, other fixed restrictions such as a maximum capital expenditure, or a space limitation. It could include fixed customer specification requirements. The rule of thumb is to minimize the boundaries unless they are true hard and fast limits that cannot be violated under any circumstances. While boundary conditions could also be included in the target section, there is value in separating the stretch goals from any absolute limitations to provide clear guidance to the participants.

Why Do We Need This Event?—This section summarizes the background and the reason why this 3P event was initiated. The explanation could include timing for product launch, market demand, profitability potential, cost targets, or any other facts that frame out the importance of why we need to develop this product/operation using the 3P process. Most companies decide to do a 3P event because they recognize the limitations of the traditional approach to design. Many of the participants will likely have been involved in those processes so focus on the positive benefits and expectations of 3P, not the failures of the previous processes. There is little point in causing participants to start off the event in a defensive mode or negative to the 3P approach.

Event Targets—In this section of the charter the more specific goals need to be stated. When must the operation begin producing? What is the maximum spend available? What functional attributes are musts? Defining *must* haves, *should* haves, and *could* haves for the 3P team is very powerful if they are already defined. This section sets the stage for the 3P event output. Note that many more criteria will be developed during the event for purposes of enabling the evaluation of different concepts, but all of the concepts delivered will eventually have to meet the target conditions to be considered successful.

When setting Event Targets it is important to give the team as much room for creativity as possible but also acknowledge the true constraints. Maximum spend, delivery timing, and location might be fixed. If they are not fixed let the team know what the estimated values are if you have them. Estimates can still be targets, and can provide good guidance to the team. There might also be certain standards that need to be met or environmental regulations or personnel safety considerations. Take some time developing good targets because they set the stage for how the team will proceed. As a rule of thumb: Big goals equal big gains. Small goals equal small gains. It is always true.

The next section of the charter simply lists the names of the people who will participate and the functions they represent, as discussed earlier:

Process Owners—They "own" the overall product or production operation and are ultimately responsible for delivering the new product and production operation. Note that the Process Owners should approve the event charter before the event. There might be two or three owners for a given event. The buck stops here.

Process Experts—They are people from every functional area that touches the product or process being created. Process experts represent groups that are involved with designing, producing, testing, or handling the product. In the most impacted functions there will likely be people at different levels of the organization ranging from operators through to Vice Presidents of Operations. This is necessary to understand and incorporate many viewpoints into the final design.

Process Outsiders—These additional people are not required for the entire span of the event but need to be ready to assist and perform work for the event team. These people need a general understanding of what you are doing, but most importantly they need to reserve some time for you in case you require it.

Judges (Critical Evaluators)—They are high-level individuals with a deep understanding of the market, business strategy, the products, the technologies, and operations. Often they are the senior decision makers. They will only participate at specific times during the event, during opening and closing meetings and for all output sessions in order to provide feedback and guidance. Because of their senior position in the company, or great experience, they will influence the group. Judges should also be asked to sign the team Contract, and everyone should understand that the entire group will be participating in the evaluation process, not just the Judges.

Event Facilitators—They drive application of the 3P process throughout the event and in some large projects might continue with the teams after the event, through delivery of the project.

Pre-Event Orientation and Training

All participants in the 3P event should meet ahead of time to go over the charter and to set expectations for their contributions. They will need to clear their schedules for the event duration. They should make plans for addressing emergencies and responding to voice mails and e-mails such that they do not interfere with the event. Participants also need to understand the physical nature of Lean 3P, working with tools and doing physical construction and assembly. Appropriate clothing and use of personal protective safety equipment will keep them comfortable and safe. A review of your company's safety policy might be appropriate for the pre-event orientation and may be repeated before any actual construction takes place during the event.

First-time participants will need more grounding in the 3P process. It is unlike anything they have ever done with respect to designing a new product or process. 3P is very intense, fun, competitive, and demanding. It is designed to push people out of their comfort zones and to challenge their thinking process. At times the process seems outright silly, and for many people this can be uncomfortable and intimidating. Knowing what to expect going into the event will help people prepare mentally and work-wise to be successful.

Additional refresher training in the Lean tools and methods may be required or advisable for some teams. There is not enough time during a 3P event to spend time training in the fundamental concepts of Lean. There need to be a number of experts on these tools in the event and all of the participants need to be grounded in the concepts and how they are applied. Foundational Lean concepts of takt, flow, and pull are musts and are nonnegotiable during the event. Therefore, it must be clear up front that these will be part of each model.

Getting Started Recap

- ☑ Foundation in Lean concepts as a prerequisite—a must
- ☑ Cross-functional team
- ☑ Event Charter
- ☑ Tools and materials for a Lean 3P event
- ☑ Space requirements

Chapter 5

The Countdown and Overview of the 3P Event Week

Timing Considerations

Planning for a 3P event will start several months before the scheduled event date. Government training grants are sometimes available to companies. If possible, it might require a much longer period in order to take advantage of that. If 3P is new to your organization, having many of the out-of-pocket costs paid may reduce the perceived risk of trying it. Although not the focus, every 3P event is a developmental experience as it brings all of the Lean tools together and applies them to a tangible project in a way that applying the methods individually cannot do.

Often the need to start the project becomes a hard driver for scheduling a 3P event sooner than the preferred schedule. When a product or new operation is announced to the organization, the pressure to go fast and deliver something creates panic and before long everyone is scurrying about to do *something*.

Have you ever witnessed this scenario? The initial development of a new product idea has been confined to a small group of people. Others were brought in to provide input into budgetary estimates and projections to determine viability with targeted sales figures, capital costs, and labor and material costs. Only a handful of people were involved because it is not certain if this project is a "go" yet, and until it is, it is on a "need to know" basis only. The potential value is significant and excitement builds as a high-level proposal is put together. It gets approved at the very next executive session, and in short order, a project manager is appointed and a team gets formed. The project cost and delivery timing are now set in stone.

Now a much larger team of people is involved from many more disciplines. Half of them are upset because they already know that things were missed in the estimates and there is not enough time to possibly get it done. There was not enough labor put into the calculations, and the resources expected to work on this project are already committed. The space they plan to utilize for the new operation has problems and will not be nearly large enough for the intended operation. It is a complete disaster and the positive aspects of creating more jobs and ensuring success of the business over the foreseeable future are all overshadowed by the weight of all these other worries.

Will 3P still work when all of this is working against you? How can you ask for a delay in starting a project just to try a new methodology for project design? Everyone knows what we are going to end up with, so why spend the time and money to do 3P?

Interestingly, this is possibly the best time to use 3P. How else are you able to deal with the probable cost overruns, the space limitations, the low-ball estimates for operating labor, and the ever-present time deadline? The need for critical thinking and creative solutions has never been more important. If doing what you did will get you what you got, then Lean 3P design becomes a must-do, not a could-do.

Taking a few months to line up the right people and get everything organized smoothly will not work. The good news is that assuming you can line up the right people to facilitate and attend the event, it is very possible to compress the timeframe and still get the full benefits of the 3P process. However, all of the critical steps still need to be completed ahead of time, so the intensity of the pre-event planning will be much greater.

In the ideal 3P timeline the following schedule and steps would be completed ahead of the scheduled event date. A 2- to 3-month delay may sound restrictive or discouraging, but recognize that product development and design are not stopping to wait. Instead the focus of the progress becomes centered on information gathering, understanding the market opportunity, the technical aspects of the product, and the voice-of-the-customer considerations. We need to shift our focus from developing solutions prematurely to gathering knowledge that will help us define the problem we are attempting to solve with our new product and operation.

As a manager, one of the toughest decisions you may ever make is to tell your project team to stop all design and development work until after the 3P event is conducted. It is contrary to their norm, and unless they have a significant understanding of 3P, their natural bias will be to get a jump on things. This will largely happen because of the pressure to deliver. Once the 3P process is established in your organization this behavior should become less of an issue.

The nature of 3P is to render thorough evaluations of risk versus reward and build safety nets for identified risks. Still, many people will be openly adverse to the proposed change and others will be silently sitting on the sidelines waiting for it to fail. As the organization gains more experience with application of 3P, most of those behaviors will lessen. Until that point, however, a healthy blend of persistence and courage may be required of a manager attempting to forge a 3P path ahead.

Preparing for a Lean 3P Event

This section addresses a timeline and activities that need to be completed prior to a Lean 3P event. While the activities all need to be completed, the time allotted to plan and execute those activities is a variable that you control. If the need to get going is critical, the timeline can be dramatically reduced as long as the key activities are completed. Small companies that are more nimble and where communication is easier might complete these steps in weeks, not months. If the urgency to start is significant, all of the steps can be executed more quickly. Do not let the timeline be an obstacle. Focus on the activities and adjust your plan for your success. Figure 5.1 graphically shows the 3P Pre-Event Timeline.

Two to Three Months Before

Arranging the event should happen early in order to get the Facilitators lined up and notify the key people and other decision makers who need to be involved. You should also be planning refresher training in Lean tools and identifying what functional areas need to be involved. Actual names of participants should be requested from the departments at this time, and it is especially important to invite the Judges (Critical Evaluators)

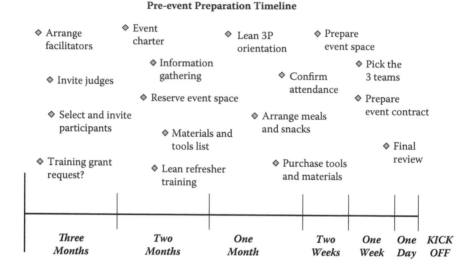

Pre-event Preparation Timeline

◆ Arrange facilitators
◆ Event charter
◆ Lean 3P orientation
◆ Prepare event space
◆ Invite judges
◆ Information gathering
◆ Confirm attendance
◆ Pick the 3 teams
◆ Reserve event space
◆ Prepare event contract
◆ Select and invite participants
◆ Arrange meals and snacks
◆ Materials and tools list
◆ Final review
◆ Training grant request?
◆ Lean refresher training
◆ Purchase tools and materials

| *Three Months* | *Two Months* | *One Month* | *Two Weeks* | *One Week* | *One Day* | *KICK OFF* |

Figure 5.1   Planning timeline for a Lean 3P event.

and get on their schedule for the week, and provide time estimates for when they will be needed. As a rule of thumb Judges need to attend the first day opening meeting and last day closing meeting and attend a 1 hour closing meeting at the end of the first day and a 2 hour session at the end of the third day. Recognize that 3P events will ebb and flow as they progress, and the timing for Judge's input will likely shift. Encourage them to carve out and dedicate larger blocks of time from their schedules if possible. If government training grants are available, they must be requested during this period.

- ☑ Arrange Facilitators
- ☑ Invite Judges
- ☑ Develop cross-functional participant list
- ☑ Invite participants
- ☑ Consider government grant request
- ☑ Plan Lean refresher training
- ☑ Plan Lean 3P orientation and schedule

Note that once Lean 3P becomes an accepted practice it can be scheduled very early in the product development process. For companies using phase gates this would be early in the Scoping phase.

One to Two Months Before

Finalize the Event Charter with the Process Owners and meet with the participants and Judges to review it. Sometimes the feedback received will suggest improvements. It is beneficial to develop the charter with your facilitators. Their experience can serve as a sounding board as you set targets and select people to participate. Your charter establishes the goals and the boundary conditions for the event and will either stretch or limit what the team can accomplish.

During this same time period you also need to develop your list of materials and tools. Every industry will have unique requirements, and having the right materials on hand for the teams to utilize makes your event much more efficient. Suggestions for materials and tools are included in the previous chapter. During the Lean 3P event we encourage participants to think like they did when they were 12 years old. This is done to remove biases and constraints, and to open up their thinking reminiscent of a time in their lives when *anything* was possible. One unique reference tool used to help the participants get into the mode of thinking like a 12-year-old are children's books that depict how things work in nature. Along with developing a list of materials required, this is a good time for teams new to 3P to do some research on available books and get them ordered to ensure delivery before the event start. Note that this point may seem slightly outrageous now but should make sense a bit later in the process discussion.

Information gathering must also be done well ahead of time. Lean 3P benefits by understanding market demand, competitive product features, and customer input or voice-of-the-customer information. Actively collecting this information in preparation for your 3P event will be critical. If key information is missing it gives you the time required to search it out.

The month before is also the right time to conduct Lean refresher training for team members who are not well versed in the methodologies. The concepts of flow, takt time, quick changeover, pull, jidoka, and poka yoke are some of the most critical as they will all be used during the design of the production operation.

A 3P orientation meeting also needs to be conducted to set the stage for what the participants new to the process can expect to experience. Experienced 3P participants might only need to cover the Event Charter, with a review of safety expectations, but people new to the process should receive additional orientation.

3P is a very powerful methodology, but it is also very hard work, requiring long days, physical and mental exertion, and open minds. The orientation meeting will help people understand how each step of the process adds value, using techniques that help them to think like 12-year-olds again and moving through the Plan, Do, Check, Act cycle over and over during the days of the event. It sets the stage for expectations and makes clear that Lean 3P Design is not like other Kaizen events they participated in. It is also a very different approach to designing a new product or new process operation.

- ☑ Finalize Event Charter with process owner approval
- ☑ Review Event Charter with Judges and Key Participants
- ☑ Reserve event space
- ☑ Develop list of materials and tools
- ☑ Obtain children's books for reference
- ☑ Information gathering—marketing data, voice of the customer, etc.
- ☑ Conduct Lean refresher training
- ☑ Conduct 3P orientation meeting with participants

One to Two Weeks Before

At this point in preparation you need to confirm attendance, order food for the participants' lunches and snacks, and arrange for purchase and delivery of the required materials and tools.

If the space being utilized for the event is available, it may be possible to tape off the three prototype areas. Be certain to review the actual space where the new operation is targeted to be located if that has been determined, and ensure that you captured the correct dimensions, including overhead restrictions. The use of photos or video will also help the participants understand the space constraints during the event.

The new production operation may need to fit into an allotted operations space, and having accurate measurements will help avoid faux pas and problematic oversights that could force you to change the design layout later. These boundary conditions need to be incorporated during the development of the three prototypes and when possible three areas should be taped off, indicating the boundaries and "monument" obstacles that are impossible or very costly to move. This applies whether you are planning full-size prototypes or scale models, and will save significant time during the event. Depict any obstacles such as columns, electrical rooms, doorways, stairways, major (unmovable) machinery, and anything else that must be considered before moving a new operation in. This task becomes even more challenging if an existing operation is in the targeted location, with plans to be dismantled or relocated. Also take note of the local sources of electric power, compressed or instrument air, potable or purified water, floor drains, and other infrastructure benefits that could influence design layout.

The 3P process requires trust among the participants. Egos need to be checked at the door along with titles, rankings, and other hierarchical positioning. Everyone needs to understand that their voice is important; their viewpoint is valued. There are no questions off limit and no one will be criticized or chastised for saying something "dumb." Some of the silliest suggestions have sometimes resulted in the biggest improvements.

To preempt this type of negative behavior a *contract* between participants is often developed and signed before or at the very start of the event. During this week before, drafting a contract covering some of key principles and making copies for each participant to sign at the event opening is a very positive step that sets the stage for how you will work together. The nature and purpose of the contract should be covered in the event orientation meeting with the participants to avoid any surprises. Some companies choose to have participants sign them a week ahead of time to allow them to think about the commitment they are making.

The *contract* may seem insignificant or even a bit patronizing, but it helps check unhealthy behavior and establishes expectations. Having participants sign the document establishes a higher-level commitment than if you just present your expectations to them. Though not a legal document, something about signing the document elicits a much deeper level of personal commitment. For some it becomes their license to open up and speak their mind or share their thoughts without fear of being embarrassed. It only takes a short time to complete but can make a huge difference in how quickly a team forms once the event begins. Many teams choose to post the signed contracts on a wall or bulletin board in the room where the 3P event will take place as a public reminder of their commitment.

The last task to consider, once the attendees are confirmed, is to divide them up into three balanced teams. The pairings should have relatively equal distributions of the key functional area representatives, considering leadership ability, technical strength, product knowledge, and individual personalities to form the teams that will work together to build the three prototypes. When there are smaller numbers of participants, there may not be enough people to split into groups of six to eight people, but for larger groups this prework will provide more thoughtful pairings and avoid confusion during the actual Lean 3P event.

- ☑ Confirm attendee participation
- ☑ Arrange meals and snacks for event
- ☑ Purchase and arrange delivery of materials and tools
- ☑ Prepare event space—identify physical boundaries
- ☑ Develop event "contract"
- ☑ Divide participants into three teams

A typical contract may look as shown in Figure 5.2.

LEAN 3P DESIGN EVENT CONTRACT

I, _____, agree to participate openly and honestly throughout this event in order to solve the problems we have facing us. I will not be judgmental or critical of other participants and will purpose to actively listen to their viewpoints and try their suggestions.

I will avoid sarcasm and conflict. I will be respectful with everyone who is on this event. I will think and act creatively and encourage creativity among the other participants by returning to my 12-year-old mindset to generate useful ideas.

I understand that our goal is to create value for our company by leveraging our combined knowledge and experience to develop the best possible product and operation within the boundary conditions outlined in our charter.

I will work with the team to ensure delivery of what is promised and committed to.

I will openly share my ideas and thoughts without fear of criticism or judgment and participate fully in every aspect of the Lean 3P Design process.

_____   _____
Participant   Process Owner

Figure 5.2   Example of a Lean 3P design event team contract.

The Day Before

On the day before the event, most of the work should already be done. If the space where the event will be held is not available up until now, it will mean setting it up and ensuring that all of the details are in place, inclusive of the materials, floor marking, Internet connections, and the many other considerations. Make sure to print out copies of the contract sheets, arrange nametags for the participants, and verify that the video projector, flip charts, and brown paper, plotter paper, or whiteboards, and markers are in place for flow charting. There will be many details to cover on the day before the event kickoff. Check them off one at a time and double check to make sure. Time is everything once the event kicks off, and nothing derails the team's progress quicker than going into a holding pattern because some critical details were missed.

If there are product samples or examples of similar products, including all packaging, literature, product descriptions, or displays available, get a number of samples for the team to use. If a product is already developed, then having reasonable quantities of the actual components will enable use of them in the prototype models to be built later in the event. If any of the components are hazardous or pose risk to the participants, use judgment in including them or pursue safer alternatives.

Along with the tactical elements of preparing for the event, the period before can be a good time to informally meet with the key decision makers or people serving as Judges, to set their expectations. It will help reinforce their commitment and ensure that the times required are on their calendars. They may have made a commitment to try Lean 3P, but until they experience the power of the process, they could be coming in with a strong conservative bias or predeterminations that will affect the 3P team. Reestablishing their commitment and addressing any concerns beforehand could be very important to the success of your event. If left unchecked, those undertones will have a huge effect on the team. Eventually they can be overcome, but it costs time and adds momentum to the emotional roller coaster that is common during this type of high-energy event.

☑ Review setup of event space
☑ Verify materials and tools
☑ Print copies of contracts
☑ Verify video projector, flip charts, markers, plotter paper, whiteboards

☑ Collect product and material samples, packaging options, literature, product descriptions, specifications, etc.

Overview of the 3P Event Week

The 3P process begins by defining the goals and criteria that will measure success in both the product and the process. To accomplish this we must first understand the product and the features or attributes that add value—features that people are willing to pay money to possess. The voice of the customer has to be considered in the very beginning before jumping into solutions. Even when there is an established product it is valuable to understand customer preferences and wants, as it provides more latitude in potential improvements. Hopefully the voice of the *internal* customers is represented within the membership of the 3P team. If that is not the case, those internal voice-of-the-customer needs should also be captured and reviewed.

Lean 3P design follows a progression that is regimented, but also intended to be flexible. Product knowledge must be developed first. Examining components, early mock-ups, competitor's product, and physical examples that are already available must be done first. The participants will physically interact with the new product, to the extent possible, to calibrate everyone's understanding. Features will be discussed and evaluated relying on voice-of-the-customer input when available. The goal of this exercise is to identify knowledge gaps. Once our understanding of the product is solid, the focus can shift to other product attributes such as packaging and display. If the product knowledge gap is significant the focus of the event will be relegated to design of the product, considering manufacturability and cost impact throughout, but as a secondary priority. Instead of operational prototypes, the teams will develop product prototype alternatives. Evaluation of the product prototypes internally and by customers will provide rapid feedback for upgrades.

Looking again at a portion of the Locher (2008) graphic (Figure 5.3) depicting the Lean 3P process, we see the relationship between product and process development. The Product Planning and Function Definition steps both need input regarding manufacturability, but they are largely focused on accumulating and processing information for the purpose of creating product value. They are precursors to the try-storming steps that follow.

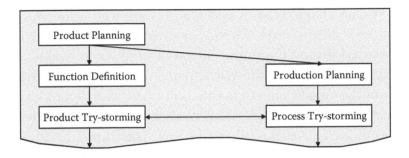

Figure 5.3 The flow of the Lean 3P process–initial phases. (Locher, D. 2008. *Value Stream Mapping for Lean Development: A How-to Guide for Streamlining Time to Market*. CRC Press, Boca Raton, FL. With permission.)

If the product is in the Product Planning or Function Definition stage of development and requires significant definition to close the information or knowledge gaps, the following process should be used first. It could be a separate event or a prelude to the weeklong 3P event. Recognize that the major information gaps defining the product must be closed before starting to create alternatives for the product and operating line. During this phase continue to consider product characteristics along with manufacturability. Table 5.1 shows product-focused information gathering steps, applying the 3P process.

Table 5.1 Product-Focused Information Gathering and Innovation

Product definition	3P event kickoff
	Sign contract commitments
	Review charter
	Affirm the high-level goals of the 3P event
	Product evaluation—components, voice of the customer, market data, competitive products, initial designs, and costs
	Define criteria for evaluating product designs
	Identify knowledge gaps
	Prototype design alternatives
	Solicit feedback and evaluate against criteria
	Second-generation prototyping
	High-level specifications, features (must haves) defined
	Select better alternatives
	Move to full 3P event

After the product characteristics and features are well understood, we then establish the critical design criteria for operations. These evaluation criteria will be used during the entire 3P event in order to evaluate the operational designs created by the participants. The 3P process then moves to an analysis of the functional steps that will take place operationally to create the intended product. Alternative methods are developed for each functional step in a creative explosion of ideas. The ideas are developed and subsequently analyzed and narrowed down to the better three ideas for each functional step. Three teams are assigned to each build a physical model of the new operation applying the sequence of alternative functional steps given them. They commence on an investigation of the technologies and practical application of the given technologies and build physical models considering takt, flow, pull, and the previously agreed criteria that affect manufacturability and costs.

In smaller companies or for smaller projects you may find that you do not have enough participants to divide into three separate teams. In that circumstance one team can develop the three prototypes one at a time, recognizing that it will take longer. Avoid reducing the number of prototype models and losing the opportunity of comparing concurrent designs, even if they are done by a single team, sequentially.

The teams along with the Judges (Critical Evaluators) evaluate the various aspects of each prototype against the agreed criteria and then select the best alternative for each value-adding step from the three prototypes to create a final mock-up that is built, evaluated, and scored. The final prototype of the proposed operation must be capable of producing the final version of the product at the targeted takt time, quality, and cost. The last step in the 3P event is to develop a plan for taking the new design forward. Table 5.2 depicts the key steps in a typical weeklong 3P event where product knowledge is already fairly well established.

Day One

The first day of the event sets the stage for the rest of the week. The participants may not know each other very well because they come from different functional areas, and operate at different levels in the organization. Quickly pulling them together as a group is important. The *kick-off* opening by a senior manager can help bring the team together by casting a vision of their common purpose. Having each participant sign the *event contract* also helps ground them and establish the criticality of teamwork.

Table 5.2 The 3P Process Steps and Target Agenda for a Five-Day Event

Day 1	3P event kickoff
	Sign contract commitments
	Review charter
	Affirm the high-level goals of the 3P event
	Product evaluation
	Product alternatives
	Define criteria for evaluating the operational designs
	Identify value-adding functions to convert materials to product
	Create high-level value stream "Process at a Glance"
	Develop seven natural alternatives for each function
	Find technically feasible applications of natural alternatives
	Evaluate the seven alternatives
	Select the better three alternatives
	Sequence the better three alternatives to develop three completely different processes
	Select three teams and assign one process to each
	Review progress with Judges
Day 2	Each team researches their process alternatives
	Develop the process elements—contact specialists, vendors, and others. Set up demonstrations if warranted.
	Physically start building the three prototypes
	Review progress with Judges
Day 3	Teams continue researching and developing their prototype models
	Teams review and incorporate as many aspects of the agreed design criteria into their prototypes as appropriate
	Teams add content to their Process at a Glance charts
	Complete the three prototypes
	With the Judges, all participants grade the merits of each prototype value-adding function using the agreed criteria
	Compare the three prototype designs function by function
	Determine the best design for each of the process functions

(continued)

Table 5.2 The 3P Process Steps and Target Agenda for a Five-Day Event (Continued)

Day 4	Identify the activities required for the report-out session and assign appropriate people to the key tasks.
	Physically construct the final prototype reusing the selected best functional alternatives from the three prototypes.
	Continue developing details of major equipment.
	Build supermarkets for materials and supplies.
	Build in Poka Yoke and total productive maintenance (TPM) elements; consider quick changeover needs and all key selection criteria.
Day 5	Finalize prototype and Process at a Glance.
	Simulate operability and material handling with the actual people performing routine functions.
	Evaluate final design against established criteria.
	Develop high-level cost estimates.
	Develop high-level resource requirements and timeline with key milestones.
	Identify highest risks from the proposed changes—and potential contingencies.
	Capture all outstanding action items and establish "what, who, and when" for each.
	Develop the plan for executing the project.
	Close out meeting with Judges.

Review the *charter* to help the participants understand the importance of the 3P event and to establish boundary conditions that have to be met in order to be successful. Everyone should have already been exposed to the charter from the earlier orientation meeting so the review should be relatively quick.

Discuss the event targets and high-level goals to set the expectations for the event. Use this review to paint a broader picture of the 3P methodology and the benefits you expect to realize.

The *product evaluation* moves the group into the more tangible aspects of the process. If available, try to have actual examples or mock-ups of the product as they exist at this time. Note that they are subject to change. If it is an assembled product, have all of the parts available. To the extent possible encourage people to touch it, talk about it, and get familiar with aspects that are of value to your customers with an explanation of *why*. Review how the product will be used by the customer, cost targets, materials being considered, packaging materials, assembly sketches, and a bill-of-materials and other information that has been developed, recognizing that it is all subject to change. Even if the event is just about adding capacity and no changes to the product are on the table, have product samples and components available.

Develop *product alternatives* comparing early prototypes, competitive products, specifications, and functions. Product features can be challenged here, comparing perceived customer value to cost. Packaging, display, and aesthetics are other areas to challenge and develop alternatives. The tension between customer value-added features and manufacturing costs should be rigorously challenged and explored across many alternatives. Prototypes and samples of the alternative approaches should be made whenever possible to increase understanding and speed the selection process. Although the intention is to determine most of the product attributes before moving the focus to process design, experience indicates that it continues throughout the entire process. This seems to be a natural extension of the collaborative and concurrent 3P process—product features that seem to be "musts" initially are allowed to be questioned when the team fully understands how difficult and costly it is to create them in an operating plant.

Establishing the *evaluation criteria* for judging the new operational designs is the next step. It is important for the participants (with affirmation of the Judges) to agree on what the evaluation criteria will be ahead of time. Later in the process there could be a tendency to bias the criteria to favor a particular approach, so agreeing upfront levels the playing field and gives each idea an equitable evaluation. We go through an extensive list of potential evaluation criteria later and discuss ways of reaching consensus on the most important ones for your project.

The first big hurdle in most 3P events comes with the step that asks the participants to identify the value-adding *functions* that take place in the conversion of the basic materials into salable product. There is a clear distinction between a process step and a function. We do not normally think in terms of function. It sometimes seems like a subtle distinction, but requiring the team to develop *functions* enables the next steps to be successful. This is the enabler that allows innovation and creativity in developing alternatives. It can only happen when you understand the essence of the core purpose or function.

Combining in sequential order the value-adding functions, we create a simple flowchart. For complex processes a fishbone diagram might be more applicable, but the requirement is to visually display the functional steps in order. This might be the simplest example of a value stream. It begins to establish the goal of achieving ideal flow in the manufacturing of the product.

Developing *seven alternatives* for each value-adding function in the process flow comes next. It is the most radical part of the process and is absolutely critical to event success. This is where you cast the big net to move people way out of their comfort zones and get them thinking differently. It tends to be very fast, very competitive, and usually very fun. During this phase of the process participants are encouraged to think like they did when they were 12-year-old children, before they knew all the technical jargon, and before they had developed solutions to many common problems. This approach tends to serve as an icebreaker allowing people who do not know each other very well to let down their guard and have some fun. It is more powerful than this, however, as it forces participants to expand their thinking beyond the constraints of the industry they know. This is done initially by having them focus on identifying natural phenomena that replicate the functional step in focus. Soon those natural methods are translated into technologically feasible approaches.

The seven technologically feasible alternatives are then developed enough to understand overall viability and risk. The team has brainstormed seven different ways to accomplish each value-adding function. Once developed the participants use their evaluation criteria to quickly rate the seven alternatives and select the *three alternatives* that are most viable. There is nothing magical about having three alternatives, but this is typically how the 3P process is practiced and it seems to work well for a number of reasons. Having several choices for each functional step of a process gives a broad but manageable number to develop and evaluate. In the next step three teams work diligently to competitively develop their alternatives, and having three teams diffuses the win–lose nature of most team competitions.

With the three alternatives selected we can very quickly arrange them into three distinct processes using the *Process at a Glance* format. Process at a Glance is essentially a process flow diagram depicting the value-adding functions (the ones for which we just developed alternatives), with additional information about the processing step, such as gauging methods, fixturing, safety and ergonomic issues, and others. Grouping the series of functional step alternatives does not require selection to combine certain ideas with others. Just line them up in order in the required processing sequence. Details of the Process at a Glance are added as the prototypes develop. Initially it will be a very simple depiction of each of the three teams' alternatives for each function displayed visually. As the models progress, more detail will be filled out on the chart. This initial version allows the teams to see a roadmap of the process they must prototype.

Assigning *three teams* after the alternatives have been developed again rearranges the group into different teams and reduces the chances of people biasing their designs based on an alternative they had earlier recommended. The pairing of people must be done carefully ensuring that each team has a good representation of technical skills, operational knowledge, and otherwise an equitable balance of functional strengths distributed to it. Also be thoughtful of the strong and dominant personalities in the group. Try to choose membership such that every team can be successful and everyone is able to contribute. Occasionally you will find someone who is very passionate about a particular idea, with a strong desire to see it explored further. In those cases weigh the benefits of moving the alternative they wish to work onto their team with the expectation that they will be open to the other alternatives during the evaluation step.

At the end of the first day the Judges should return and be taken through a summary of what was covered. This meeting may take an hour or longer depending on how much change has been recommended to the product or process. The first day covers a lot of 3P process steps, but many of them go quickly. Plan on working a long day in order to get through everything, inclusive of the wrap up with the Judges. Alternatively, the wrap-up session can be scheduled for the next morning. Avoid this if possible, due to the amount of work that lies ahead.

Day Two

The next day is spent with the three teams working together to develop their ideas and prototypes. They will be researching aspects of their designs where the technology or approach is unfamiliar. Vendors may be called in or visits to key suppliers may be arranged. This is a time for each team to combine all of their skills and knowledge to create a complete operation that incorporates all of the assigned alternative functions.

At the end of Day Two it is common to bring the Judges back and allow them to walk around visiting each of the three teams as they evolve their ideas, thinking, and prototypes. This is especially valuable for companies new to the 3P process. There are no formal presentations. This time helps the Judges understand the selected alternatives better and is very affirming for the three teams receiving their support.

Day Three

On Day Three the teams continue to research and complete their prototypes. It is not uncommon to have vendor visits on the third day or to have
representatives from a team visit a supplier if they are located within a short
driving distance. This part of the process is about technical viability and
finding commercially workable solutions adoptable on the scale of the operation being developed. Beyond the functional process step alternatives, each
team will develop transport or transition steps to connect the processing
functions. Many types of gauging will be used to verify that the intended
process function was done correctly. Takt, flow, pull, and other elements of
Lean will be built into the prototypes.

Once the *three prototypes* have been physically built, the teams and the
Judges evaluate the designs. One team presents and the other two teams
and the Judges ask questions, challenge the basis for their designs, and listen
to the team's assessment of their design's performance against the criteria.
Every functional step is evaluated and rated against the preselected design
criteria and event targets. An overall assessment of the team's approach is
also discussed and noted. The other two teams follow immediately afterward in similar manner.

After all three prototype presentations, the participants with the Judges compare each alternative function from all of the three prototypes, comparing features, risks, and benefits, along with the evaluation criteria ratings. For each of
the three alternatives, a best choice is selected. The combination of the selected
alternatives forms the input for a final hybrid prototype that will include the
very best ideas at every step of the process. Sometimes a team has an overall
approach that has distinct advantages and will become a feature of the new
operation. Transitions where product moves from one value-added step to the
next are often selected from a model as a best choice, even though it was not a
focus of the evaluations. The goal is to take the very best ideas from the entire
team and amalgamate them into the final design.

Day Four

The three prototype teams are dissolved and instead there are new teams
formed to develop various elements for the final report-out session. With the
final modules for the *combined prototype* selected, the participants generally
work together initially to physically take apart the three mock-ups and begin

combining them in one of the workspaces. A new Process at a Glance is developed and will be added to as the final days ensue.

There is more focus on incorporating the other Lean elements at this juncture and optimizing the total operation. Supermarkets, quick change-over considerations, total productive maintenance, Jidoka (the ability of a machine to stop and provide notification of a problem), Poka Yoke (mistake-proofing), takt time estimates, and line balancing now need to be incorporated into the model. Additional detail is considered such as material flow to and from the proposed process, and the capability of supplier processes for purchased materials.

Day Five

While some members of the group continue to complete the physical model and update the Process at a Glance, other teams are split off to develop the cost estimates and preliminary resource plans and timelines along with key project milestones. *Project resource plans* will differ from most traditional plans because they will include cross-functional resources from the members participating in the 3P event. This is not a casual change, but a very dramatic one that the organization needs to be prepared to support. It will require some time from the entire cross-functional team of participants as they work with Engineering to develop the operation, typically with no less than weekly meetings to update them on progress and to work through any issues.

The final step of the Lean 3P Design event is the *report-out* meeting. It could take 2 hours or more to go through the final report-out. Included will be a review of the physical prototype, Process at a Glance, evaluation against the criteria, targeted goals and boundary conditions, and the project and resource plans. The last day should also incorporate a "lessons learned," capturing what worked well and what needs improvement the next time. There must also be a time of celebration and recognition for the brilliant designs and hard work accomplished by the team.

The next chapters progress through each Lean 3P process step in more detail in order to expand on the content with a focus on how and why each aspect of the event is important and additive to the whole process.

Chapter 6

Kickoff, Charter, and 3P Goals

Kickoff

At the start of the 3P event it is useful to have the head of the organization or other senior person kick it off with an opening greeting that sets the tone and expectations for the event. It does not need to be long, but a sincere statement of expectations and confidence in the desired outcome can be empowering to a team new to 3P.

Introduce the Facilitators and the people who have agreed to be Judges (Critical Evaluators) for the event. Because it is common to have the Facilitators come from outside of the company or perhaps another company location or business unit, introduce them and turn control of the event over to them. It is worth recognizing their expertise in Lean and specifically in 3P. The 3P process is different than what most participants have seen before, and they must trust the Facilitators to guide them through the inevitable rough patches they will encounter during the week.

Introduction of the Judges recognizes them as senior people in the organization and allows the team to see their endorsement of the process. If done well, these introductions also demonstrate the confidence the company has placed in the participants to deliver something very special. Explain the role of the Judges—to provide strategic and deep technical insight—and what the role is not—making final decisions on the designs developed by the teams.

Some organizations like to start group events with a "go round the table" introduction. Participants are asked to state their name, expectations, and experience as a type of icebreaker. Because the participants will likely come from all around the organization and even outside of the immediate organization an introduction is warranted. How much more you choose to include is probably dictated more by company culture versus any specific benefit to the 3P process. Although 3P is always developmental in nature, it is not the driving purpose for having a 3P event. The purpose of the event is entirely about creating an effective design for a new product and a new operation to produce it. In situations where there is already an established product and the event is about designing a new operation, the same focus applies.

The 3P Event Charter

Reviewing the Event Charter is the beginning of the real work. An example of an Event Charter can be seen in Chapter 4 (Figure 4.2), and a template is provided in the Appendix. The *Process Name* might be intuitive so just state what it is unless there is something significant about it that the participants should understand. For example, the operation being designed may be specific to one product at this operating plant; however, the name could reflect the intent to take whatever is created and eventually standardize the design across multiple sites. It will not change the nature of the event, but it could help participants understand the significance of what they are doing and their impact on the larger enterprise.

The *Process Boundaries* must clearly state what is in and what is out of scope for the event. Where does the process start? Where does it end? Let the team know if they have latitude to change any aspects of the product or the packaging, being careful to not prematurely quench potentially great ideas by doing so. A value stream map could possibly be provided that goes from the supplier of raw materials through manufacturing, shipping, and into the customer's hands. Every event needs a very defined starting point and ending point between which they are expected to be highly creative in the development of their design. Outside of that space they need to know they have no jurisdiction beyond understanding external constraints and interfaces.

As mentioned earlier, other boundary conditions can be included in this section of the charter. Sometimes when the teams begin defining the value-adding functions great questions come up that can step over the boundaries. "What if" can sometimes lead to one of those amazing "eureka moments," redefining how things are currently done with the potential for great benefit. The facilitator or Process Owner must decide to allow a new idea on the table to be pursued even though it violates a boundary condition. The answer might be a qualified "yes." Progress the new idea as quickly as possible to get a response from the necessary stakeholders, but concurrently stay the course on the design development within the defined boundaries assuming that the alternate might need more time to be approved than the project allows. This guideline is particularly true relative to product design.

Take some time to focus in on the *Why We Need This Event* section. The charter is intended to be succinct, but when going over it with the group there is opportunity to embellish on the key messages in order to increase understanding and motivate the team. Give some background as to the value of the project, expected revenue, benefits of the product, competitive pressures, cost or quality issues, and other pertinent information. The "why" should also express the reasons for the choice of 3P versus the normal approach used to design a new operation. Give examples of previous successes you would like to emulate or of problems you need to avoid. Most people appreciate the limitations of the traditional design process and the broad range of participants will quickly understand the value of having their voice considered in the design of the new operation.

The most common statement made to me at every 3P event I participated in is "This is the first time I've ever been asked to contribute to the design of a new operation or provide input into a new product." That statement is always followed by, "They never ask us until after the new operation is running and the problems have occurred." Do not dwell on the negative aspects of previous design efforts. Someone sitting in the room with you (or their departments) likely created most of the negative examples you would be inclined to use, and there is no need to build up defensive walls for the sake of making a point. Keep it positive, always pointing to the expected value creation.

The *Event Targets* should be covered in enough depth to familiarize everyone with the goals that must be achieved. Good targets help the participants realize how Lean concepts must be applied for them to be successful. Stretch targets set the bar high and participants will realize that attaining them using the normal design process is not possible. They should consider basic information like customer requirements, known specifications, and market demand. Much of that needed information will be provided as an output from other elements of the Lean Product Development process. Event Targets can include the maximum money available to spend on the operation, location requirements, maximum number of operators, and almost always a timeline. Do not invent these if they are not required, as it will give the teams more latitude to be creative. There can also be implied targets that may not be absolute but are strongly desired. Differentiate between the two. This is also a place to develop the *mandatory* design input requirements (DIRs). Draft DIRs should be an output from the event after the final prototype is built and the criteria evaluated. However, at the start there might be some mandates that the team has to consider. Takt time, process yields, throughput numbers, and changeover times can all be included as Event Targets if appropriate. A lot of consideration should have been put into this earlier. Like the process boundaries, the Event Targets specify the critical criteria that teams must achieve. This is not the time to be adding targets. That phase should have been considered and agreed to long before the start of the event. When reviewing the Event Targets ensure that all of the participants are well grounded in the concepts and how they will be applied and measured.

The *Event Schedule* sets the expectations for when the participants need to be involved. 3P is very intense, and while longer days may be necessary and desired, remember that people will be working very hard, mentally and physically. It can be stressful at times and people will need breaks to refresh themselves and to catch up on phone calls and e-mails from their "day jobs." If you do not specify when they are allowed to do those activities, people will take their own initiative and start to wander off.

For a 5-day event, starting at 8 A.M. and going to 5 P.M. usually works fairly well. Working lunches can also help the teams stay focused while they physically recover. Having good food, coffee, and drinks available for them lightens the event and helps productivity. Tell them when breaks will be taken. Let them know that nonemergency calls should not be taken and doing e-mail needs to wait until designated times. Some groups allow for formal breaks at specific times and state times for checking messages. Do what makes sense for your organization, but the clear goal is to keep people focused on the long week ahead. Facilitators will need to plan for extra time with key people at the end of each day to do a review of the day's activities and make adjustments for the following day.

Union facilities may need to consider any additional issues that could impact how the participants get chosen and what schedule they will work. These are not insurmountable but must be considered and worked through in order to get the needed benefits of their input. In all cases nonexempt personnel will need to be compensated for total hours worked, and these need to be built into the plan when pursuing working lunches or extended

Lean 3P Event Timeline

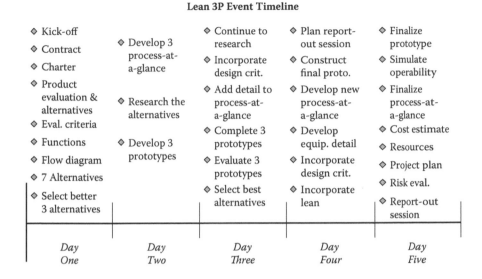

Day One	Day Two	Day Three	Day Four	Day Five
◇ Kick-off	◇ Develop 3 process-at-a-glance	◇ Continue to research	◇ Plan report-out session	◇ Finalize prototype
◇ Contract		◇ Incorporate design crit.	◇ Construct final proto.	◇ Simulate operability
◇ Charter				
◇ Product evaluation & alternatives	◇ Research the alternatives	◇ Add detail to process-at-a-glance	◇ Develop new process-at-a-glance	◇ Finalize process-at-a-glance
◇ Eval. criteria		◇ Complete 3 prototypes	◇ Develop equip. detail	◇ Cost estimate
◇ Functions	◇ Develop 3 prototypes	◇ Evaluate 3 prototypes	◇ Incorporate design crit.	◇ Resources
◇ Flow diagram				◇ Project plan
◇ 7 Alternatives		◇ Select best alternatives	◇ Incorporate lean	◇ Risk eval.
◇ Select better 3 alternatives				◇ Report-out session

Figure 6.1   Timing of activities in a five-day Lean 3P event.

hours, especially if it will result in overtime pay. Every operation handles management of overtime differently, and this should be thought about ahead of time to avoid possible distractions during the event. Figure 6.1 shows the typical week-long Lean 3P Event Timeline.

When covering the list of *Event Participants*, help everyone understand why each person is there, what aspects of the design you need them to contribute most to, and why they were selected for the assignment. Often the reason a specific individual was selected was due to their underlying Lean experience or because they also worked in Operations or another key department in addition to the area they now work. In every case they are expected to be subject matter experts in the functional area they represent.

Event Participants include Process Owners, Process Experts, Process Outsiders, Judges (Critical Evaluators), and Facilitators. Details of their respective roles and suggestions for people to include were mentioned in Chapter 4.

Ensure that everyone has the opportunity to comment on the charter and even debate some of the points if they need to. Hopefully it is not the first time that they have seen the charter, but it might be the first time they have been able to question it. Do not take too much time doing this, but be careful not to shut anyone down, as the group will be watching for signals to see how open you will allow them to be.

Introducing Lean 3P Design Goals

Spending a short amount of time regrounding the team in Lean 3P Design will happen next. For some it will be redundant, but for a number of people it will help them see where the event will be going during the week.

3P is a process to develop a Lean production "system" in the shortest time possible that creates the right product with the right features in a producible product design that

- Produces the highest *Quality* parts
- Produces at the lowest possible *Cost*
- Produces product at the required volume per the *Takt Time*
- Produces with the shortest necessary *Lead Time*
- Meets customer expectations

The team should be challenged with aggressive targets. The 4-¼-4 challenge mentioned in the Introduction represents the kind of targets a 3P event should strive for. Can we design a process made up of four smaller operations that share the demand requirements? Can we minimize capital spending for those four operations and keep the total cost at one-quarter of what it would cost to build a single larger line? Can the ongoing costs to operate and maintain the four smaller operations give us four times the productivity, reflected in our standard costs per part produced? Recognize that the 4-¼-4 challenge is not a formula or an edict. It is simply intended to stretch our thinking—to force us to radically look beyond our paradigms, to a place that initially might appear impossible. With this starting point, the imaginations of the event participants can go to work and begin collaboratively exploring "what if" possibilities.

To achieve step-change benefits we need a producible product and a production operation with high yields and utility. This requires designing out wastes from the start and ensuring that product requirements and specifications are adding value and the manufacturing operation can consistently meet them. It might require eliminating product attributes that are not adding value. Walmart did this when they began requiring suppliers to provide deodorants and other bottled materials to their shelves without the cardboard boxes that had been the standard in that industry. They recognized that the added packaging material added no value but did add cost and had a negative impact on the environment.

Most production operations are highly complex. Typical operations include processing steps for altering raw materials by cutting, combining, or forming them into something with a higher value. Many operations then purify, clean, or process them into a final form that gets packaged or assembled, labeled, and finally prepared for shipment to a distribution center and the ultimate customers. With 3P all of those processing steps ideally meet a targeted takt time and are joined together by synchronized flow.

Takt time equals the available time per day, divided by the daily customer demand, often expressed as seconds per piece. Controlling takt in every section of the operation helps ensure that project costs are maintained at the lowest possible level because you are not wasting money by overbuilding and will not be at risk of underdelivering. Too often engineers focus on the key processes only and overlook the peripheral functions, like using a vision system to scan a product label, only to find that the scan and vision analysis take longer than the value-adding step. These details must be made visible throughout the full development of the operating line. We need to balance the takt time of each process step including transition steps and gauging. We also need to look at other constraining processes that could impact takt time. Examples of this could include quality sample analysis of the first product of a new lot, or replacement of consumed materials such as printer ribbons, or time-sensitive operator paperwork. These peripheral processes all detract from operational output and should be evaluated.

Takt time should be calculated before the event, recognizing that it will be an estimate based on a forecast demand. However, there are benefits in walking the team through the calculation. Most businesses have uneven product demand. No matter how well we forecast or what model we use, customers somehow have a way of doing what they want to, and even our best forecast will be wrong. It is only a question of by how much. A sales demand forecast over time (including surges and slow periods) is a necessary input to the takt calculation. Make-to-order products and demand-driven pull systems rely less on forecast accuracy, but there still must be predictions made about future demand to ensure the necessary production capacity.

The takt time calculation is relatively straightforward. If demand is seasonal, use the higher months or quarter estimates to base your takt time. Level the demand to minimize the impact of momentary spikes to avoid overbuilding, but do look out over the appropriate time horizon for your business, balancing the uncertainty of the forecast against the time and cost to add more capacity.

TAKT TIME CALCULATION

$$\text{Takt Time} = \frac{\text{Available Time per day (seconds)}}{\text{Average Customer Demand per day (pieces)}}$$

■ Available Time reflects the amount of time in a day that the operation will operate, making product. Shift changes, lunch breaks, change-over losses, equipment outages, and micro-stops must be subtracted from the available hours if production rate is impacted.
■ Average Customer Demand per day, with assumptions for surges, seasonal demands, and growth rate expectations.

Once those determinations have been made, takt time can be calculated. State the quantity of product required per year, quarter, or month. How many days per week will the operation run? How many shifts? Sometimes the number of days or shifts can be varied to handle occasional surges in demand or seasonal inventory builds. Estimate the number of hours per shift the plant will be producing good quality product. Include start-up and shut-down losses, preventative maintenance activities, breaks or meetings that take labor away, holidays, plant shutdowns, and the anticipated process capability for the operation based on other similar operations, if available. An 8-hour shift will often only result in 5 to 6 hours of production time, depending on these variables. Note that this discussion can be enlightening and controversial especially with stakeholders who are not familiar with typical production metrics. Do not be too aggressive when estimating process capability or effective operating hours. It is best to assume that past performance is a good indicator of future results, unless the improvement desired is a stated goal in the 3P Event Charter.

Taking the available production time per day and dividing it by the quantity needed per day will give you takt time, in seconds per production unit. All 3P events should have takt time as a criteria, and the challenge of building each value-adding functional step to meet the takt pace helps ensure flow, eliminates work-in-process (WIP), and ensures the lowest possible capital cost for the operation. Consider how this might apply to our 3P design. Let's use an example where the overall takt time for a given product is 5 seconds per part. We can design a single line to produce it at a certain cost. What if we were to design it to be manufactured on four lines versus one? Now we would have takt time of 20 seconds per part, using four operating lines. Producing a part every 5 seconds might require a very different approach than a part every 20 seconds. The equipment would probably be bigger and might need to be more accurate to support the faster rates. A slower rate might permit different approaches to be used that would not work at the higher rate. Many new possibilities open up for consideration when you start by understanding the anticipated takt time.

To achieve the lowest costs, the principles of *Flow and Pull* must be a first choice for incorporation into the operation's design. Flow ensures that all steps of the operation feed seamlessly from one to another with minimal distances and stopping points. If there have to be stopping points between batch operations, develop Pull systems to trigger resupply. Avoid or minimize buffer inventory between steps to reduce WIP inventory with all of its inefficiencies. Ideal systems produce one part at a time through a system that simply flows all the way from start to finish with perfect yields and quality every time. Real-world systems are lumpy. Machines break or have micro-stops. Materials have natural variation that must be accounted for. Different departments handle different functions and steps, often without standardization. Space considerations do not allow for adjacency in layout. Understanding these limitations at the start of the design process allows for optimization that will yield benefits over the life of the product.

Failing to consider Flow and Pull in the overall design will destine you to a substandard design. Understanding and prioritizing Flow and Pull at every step will enable the teams to think about what the real constraints are and how they could improve them one aspect at a time.

Error proofing or *Poka Yoke* is another significant design consideration. Attaining the highest possible product quality means devoting a lot of thinking to making sure the process is capable of producing every part reliably and consistently. The product design should consider using fewer parts, and parts that can only be assembled one way. Specifications must meet quality requirements and must also be readily achievable. The operating equipment should be reliable and robust and built to be easily maintained. The goal is to create a product and process that makes it *difficult* to produce bad parts. Designing with this attribute in mind is essential but not common because we tend to think of the ways things *should* work, focusing on getting the requisite functional operation completed so that we can move onto the next. We do not normally think about the failure modes at this level and build in the robustness to either the product or the process that prevents the failure from happening in the first place. This is where the experience of operators, maintenance personnel, and engineers who have wrestled with this type of variation in the past will be critical during the prototyping stage, when these aspects of design are being considered.

The concept of designing quality into the whole system is paramount. We do not want to simply inspect quality by applying numerous quality control checks or by adding expensive vision systems, sensors, or instrumentation along every step of the way. Some of those elements will be necessary, but how much better is it to make the product and process so robust that it becomes virtually impossible to make a bad part? Designing quality in is the goal.

Cycle Time looks at the time it takes to produce one part or unit of product in one cycle, from start to end. It is a factor in determining overall capacity and WIP. Cycle time works in relationship with takt time, but they have a different focus. Two plants that both produce at a takt time of 20 seconds per part might have very different cycle times. If one has a 30-minute cycle time and the other has a 90-minute cycle time, the amount of equipment required might be vastly different. Certainly the amount of WIP in each operation will be very different.

Not all industries use the same definition of cycle time, and it must be defined clearly if it is used as an event target. With batch processing, cycle time usually reflects the bottleneck step or part of the system that constrains the rest. In chemical batch processing where product moves from tank to tank through various reaction and purification steps, keeping track of the longest cycle and the amount of material being processed in that step is

Overall equipment effectiveness (OEE) is an increasingly common statistical measure that combines several individual metrics into one. It is excellent for monitoring performance of an existing operation. Generally it is a *poor* metric to use in 3P because of its complexity and specificity to a particular operation. Quality rate, run time, and run rate metrics are used to generate OEE. OEE multiplies the percent of good quality parts times the percent of total planned time the operation was actually running and producing product, times the percentage of actual run rate versus the intended design run rate. With three levers influencing the metric, OEE can be manipulated many ways. Ultimately the individual components of OEE must be improved to affect performance of the operation, and it is typically more effective targeting those in the event instead of the combined metric.

critical to ensuring demand will be met. Meeting takt time in each of those steps would require balancing the time per unit of output at *every* processing stage. When using cycle time as an event goal, it is encouraged to use a definition reflecting start of materials in and completion of final product out.

In continuous process operations, start-to-finish cycle time correlates directly to WIP, and reducing cycle time will have a direct impact on operating costs. In discrete manufacturing, cycle time is similar to batch but normally reflects the time from when a part enters to when it comes off the line. Though not optimal, sometimes discrete manufacturing is done in separate operations with inventory buffers between major steps. Each of the separate operations might meet takt, but monitoring total cycle time and finding ways of reducing it will reduce in-process inventory and operating costs.

Each Part, Each Interval (EPE or EPE I)

Companies who manufacture many different products on the same operating equipment can use *each part, each interval* (EPE) to determine how often a product can be made during a given interval without exceeding capacity. Operational capacity takes into account takt time calculations in conjunction with EPE. This type of manufacture is especially challenging due to the probability that the multiple products will have different processing steps, requiring different amounts of time, and will often all have separate sales forecasts. These differences make optimization of a new operation or accommodation of a new product a very complex problem. It becomes critical to understand and optimize lot or batch sizes. Changeover analysis also becomes a huge factor for consideration. Sometimes different products can be grouped into similar *families* of products where segregation or campaigning of products can provide advantages.

Summary of the 3P Design Goals

During the opening of the 3P event the relevant concepts should be discussed with the participants, focusing on what is important to your business and relevant to your operation. Time to market, capital cost, takt time to satisfy customer demand, and product cost might be hard and fast boundaries that absolutely have to be achieved. Some of these elements, like EPE, are for consideration and are not necessarily appropriate for

Example of 3P Event Goals

Event Goals and Targets
Standard product cost under $5.35, at rate of 240,000 units per year
Time to market launch 14 months
Operational hours - 3 shifts, 5 days per week
Must achieve takt time = 64 seconds per unit produced
Must incorporate flow, pull and Poka Yoke
Capital cost limited to $350,000 USD
Capital cost stretch target of $150,000 USD
Cycle time under 30 minutes is desirable

Musts: Standard product cost under $5.35, at rate of 240,000 units per year; Time to market launch 14 months; Operational hours - 3 shifts, 5 days per week; Must achieve takt time = 64 seconds per unit produced; Must incorporate flow, pull and Poka Yoke; Capital cost limited to $350,000 USD

Wants: Capital cost stretch target of $150,000 USD; Cycle time under 30 minutes is desirable

Figure 6.2 Differentiate "must" and "should" goals.

your project. Some companies include stretch goals or challenges for the 3P team. Be clear which goals *must* be met versus others that are challenge targets. Figure 6.2 demonstrates the differentiation of "must" goals versus "should" goals.

In the next chapter we will go over design evaluation criteria, which are additional areas of emphasis that the participants will select in order to compare the relative strengths and weaknesses of their alternate designs. Some of them will also be must-do's. The difference between those and the event design goals included here is hierarchical, with the higher importance on the event goals that appear on the charter. With awareness of the relative importance of the goals, teams can go to work achieving them.

Kickoff, Charter, and 3P Goals Recap

- ☑ The kickoff, introductions, and setting high expectations
- ☑ The role of the Judges
- ☑ Introduce the Event Charter
- ☑ Typical event timeline
- ☑ Event design goals
- ☑ Present which design goals are mandatory
- ☑ The importance and calculation of takt time

Chapter 7

Product Planning and Process Evaluation Criteria

Product Planning

Developing a new product should be a marriage between a genuine customer need and your capability to provide a solution at a price point that creates value for the customer and meets your profitability targets. That simple statement underscores the myriad of complexities that immediately begin to enter in as soon as we begin to consider a tangible product for real customers.

Understanding the needs and wants of the customer and meshing those with the core competencies of the business and manufacturing organization is the challenge every new product faces. A product evaluation should be done to understand the attributes and features that have perceived value to a customer. Market data, voice-of-the-customer data and competitive products information are key inputs, often coming from other events in the Lean product development process. On the Operations side, manufacturability and materials costs are the key variables. Figure 7.1 shows the "sweet spot" where the optimum customer product intersects with the optimum manufacturability.

As we start to evaluate the product, we have to question what differentiates it from competitive products. Can we list the key differentiators that the product provides better than competitive products or previous solutions? Can those attributes be prioritized and rated for customer perceived value? Does customer information support our ratings or does it need to be vetted?

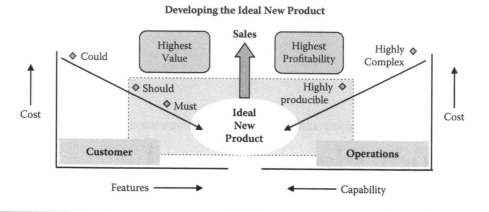

Figure 7.1   The ideal new product balances product features with manufacturing.

Now consider manufacturability. How complex is the product? Does it take advantage of your core competencies or is there high technical risk? Can it readily be designed for manufacturability or assembly, or is it highly complex and expensive to build? Are the material costs able to meet profitability expectations as well as quality requirements, or will alternate materials need to be explored?

A chart like the one in Figure 7.2 may help in your evaluation of product features or differentiating attributes. List the known features that you know the customer wants. Brainstorm to generate other possible product features with your team might have. As you talk through the various features consider how well they align with your core competencies. Does your manufacturing organization know how to do the manufacturing or assembly of the product with the proposed feature? Consider the complexity of manufacture and how easily it might be simplified to make production of it robust. Rate the product feature on a scale of 1 to 3 for Importance, based on perceived customer value. Then rate the feature on a scale of 1 to 3 for Manufacturability. Add the scores to get a total score for the feature. Look at the highest scores. These are the features that should have the greatest potential of pleasing your customer and should be the most profitable based on relative ease of manufacture. If possible, construct prototypes of the highest-scoring product functions and physically evaluate them. Make adjustments for improvement. Get real customer feedback where practical. Figure 7.2 uses the example of a remote controlled model car.

Utilize the feedback from the Product Evaluation to develop a preliminary design for the new product. The balance now transfers over to the process side of the design effort.

Product Evaluation

	Attribute/Feature	Importance Must (3) Should (2) Could (1)	Manufacturing Difficulty Low (3) Medium (2) High (1)	Total	Consider Rapid Prototype Mock-Ups
1	One-hand Remote Control	3	2	5	
2	Rechargeable Battery	2	3	5	
3	Customizable Decals	1	1	2	
4	Unibody Design	2	3	5	
5	User Repair Kit Inclued	1	2	3	
6	Pro-pack Accessory	1	2	3	
7	Super-grip Tires	3	3	6	
8	Real-rubber Smell Kit	1	1	2	
9	Extra-loud Rumble Woofer	2	1	3	
10	Voice Command	1	1	2	

Figure 7.2 Evaluating product features and manufacturability to find the optimum balance.

Production Evaluation Criteria

Determining the criteria that will be used to evaluate the process design at the outset enables the team to establish unbiased metrics. Everyone will have some biases initially, and most will begin to develop stronger biases as they work with their teams progressing the various design concepts. 3P (Production Preparation Process) tends to be very competitive, and the teams will get excited about the possibilities of the new ideas they develop. It is easy to be swayed by those new ideas and end up force fitting the criteria to preferentially select them. Although the 3P process continues to realign people as the event progresses so that the ideas they develop at one stage are often progressed by someone else at the next stage, determining what criteria are important up front minimizes the impact of bias and helps set expectations and direction.

The 3P participants and Judges (Critical Evaluators) will determine the criteria to be used for evaluation considering all viewpoints and working to understand the impact of the criteria on the total project and the people who will interface with it. This selection process forces the team to rate the relative importance of each criterion and choose the ones that will eventually be used to judge the success of each design proposal.

The following section suggests 30 possible design evaluation criteria to choose from. Some will undoubtedly be very familiar and others may not. Some may be more specific to industries outside of your own. As the team evaluates the relative importance of these criteria some will be discounted as minor or insignificant to your type of operation. Select the criteria that you believe will have the most benefit. There are different ways to prioritize criteria, but a brief discussion may be enough to render a thumbs-up or thumbs-down decision. Some groups prefer a simple weighting factor system. Another simple way to evaluate the relative importance of design criteria is to ask the question: Is this something we *must* include in our design, *should* include, or *could* include? Framing the discussion using this set of priorities may help to focus the group on the most important. A checklist to assist in this determination is included in the Appendix ("Potential Design Evaluation Criteria Checklist").

3P Process Design Criteria

Takt Time

Takt time refers to the pace of an operation, with the goal of producing no more and no less than customer demand requires. Takt is derived from the German expression for drumbeat, and conjures up images of a conductor waving a baton to direct an orchestra. Here it is used to establish synchronized production, where each step of the process is scaled to make and pull materials from one station to the next in the same amount of time, ideally without the need for inventory buffers in between. Each step in the manufacturing process is designed to meet the required customer takt time. Takt is almost always a criterion for 3P because it is so foundational to design of the process.

Many times takt will be based on speculative data, and rough estimates will be the best you have to work with. In those cases consider using 10× multipliers to provide the scope for process design. What would the operation look like at 1000 seconds per piece, or 100 seconds per piece, or 10 seconds per piece? Chances are that there will be significant differences. Note that the more speculative the takt time number, the greater will be the value in developing designs with multiple lines with low capital cost, that each do a proportional percentage of the demand requirements. Additional lines can then be built as the demand increases. Please see a further explanation of takt in Chapter 5.

One-Piece Flow

One-piece flow means moving one part or workpiece at a time through the production process in a system that flows from start to finish. The goal is to meet takt time and to avoid inventory buffers and stops between operating steps. One-piece flow applies to discrete manufacturing of single parts as in a Job Shop operation, with the challenges of uneven demand and high mix. It works in more commoditized operations. We should not take the "one-piece" terminology too literally. In batch operations it could mean producing multiple small batches versus a single large one. The approach of reducing batch size and reducing changeover time can be referred to as economy-of-repetition, versus one-piece flow, however the concept is the same.

One-piece flow can apply to continuous plants by choosing to build multiple smaller lines versus one large one that meets the total demand requirements. The challenge with this element is twofold. First, do not falsely assume that you are not Lean without literal one-piece flow. Lean is about continually improving efficiencies and a batch of one does not work for many product types. Even in operations designed for one-piece flow there can be places that must be augmented using Pull systems. Second, consider and understand the trade-offs so that they are not dismissed as irrelevant in traditional batch or large commodity operations. Even large-scale continuous process operations often have elements of batch production someplace in the line. There can be great benefits in finding the sweet spot among the trade-offs.

Pull System

The concept of Pull requires that we do not overproduce, but wait until a downstream process tells us to make the next increment of production. This is counter to a Push system where each part of a process is suboptimized to produce a defined batch size, or at a higher rate, building an inventory buffer that gets pushed out to be accumulated until the next process can start to work on it. In Pull systems, a downstream process creates a signal for an upstream process to produce the next part or lot of material and only the required amount is produced. Pull eliminates the need for inventory buffers between process steps, except to balance the line.

Pull can apply to the resupply of raw materials or other supplies used during routine operation as well as between the value-adding process steps. Though every design should consider ways of minimizing inventory buffers, a focus on ensuring an effective pull system with pull signals and good flow might be very beneficial. A process that creates a pull system assumes that some batch processes are necessary and rewards designs that meter product from station to station in a manner that minimizes or eliminates inventory buffers. When done well, this internal pull system drives down inventory and other wastes associated with overbuilding.

At its best, Pull starts when a customer places an order or withdraws product from a distribution center, and a signal triggers manufacturing to build the amount required. The Pull signal continues to spark new demand requirements until eventually the raw material suppliers are sent notice to resupply. If the material inputs or related functional activities (such as laboratory analysis) are considered within scope of your 3P event, remember to include them within the boundaries section of your 3P Charter. With such long supply chains in place, even the best Pull systems typically have inventory buffers associated with them. Comparing Pull to the more common batch process planning, Pull systems require more understanding initially but yield significant benefits over batch planning when thoughtfully implemented.

Many companies utilize enterprise resource planning (ERP) software to connect and manage key areas of their business, including finance, customer orders, inventory, quality, maintenance, manufacturing, logistics, and many other functions. Most ERP systems are capable of supporting material requirements planning (MRP) or the next evolution manufacturing resource planning (MRP II). These processes are typically very intense and require strict adherence to key process metrics to be successful, and at their best, they require a demand plan that understands what customers will order. There is much written on their application and limitations, however MRP II is essentially a push system. Developing Pull within an MRP II environment, even if it is only relegated to a small portion of the products (typically the products with the largest volume), can have huge benefits. It is common to have both MRP II and Pull being applied together, enabling both processes to do what they do best. Incorporating Pull at the ERP level might be beyond the scope of your 3P event, but it is a good factor to consider in postevent activities.

People Involvement

People Involvement evaluates the level of people interaction and intervention in a smooth process flow. It seeks to assess the balance between people contributions in the performance of standard work, versus automation both in terms of costs (initial and ongoing) and flexibility. The team choosing this criterion needs to develop the definition of the metric to ensure common understanding. Is the intent to minimize labor? Is the goal to optimize the balance between higher-cost automated designs versus higher ongoing costs and more people dependence? Flexibility can also be a factor especially when the same operating people must support other operations, too. In this case, standardization might weigh heavily in the design, making the operator interfaces very similar to existing operations. If quick changeover is a criterion for the process being developed, the team should also consider the people involvement for this aspect of the process. Quick changeover often involves drawing on other people during the changeover downtime period to expedite the required activities. The team should consider where those extra people will come from and what will trigger their support for the short periods of time required.

Having operators, material handlers, and maintenance people involved in the 3P process helps ensure that these elements are considered. Selecting People Involvement as a design criterion will force specific consideration of these features to be evaluated. One consideration when selecting these important participants is their ability to confidently contribute in a group heavily biased with very technical people, without being intimidated. Their role is equally important to a successful design.

Automatic Unloading

Automatic Unloading refers to a feature built into the operation that automatically processes the last part through to always leave the machine available for the next run. It can reduce confusion and errors and decrease setup time. It is usually a low-cost option to include relative to other automation schemes.

In most cases Automated Unloading is software controlled, requiring a sequence to be triggered and completed. In automated systems, consideration of the impact of unintended stoppages is typically built in to protect the equipment, people, or product integrity. Unintended or triggered stoppages of automated equipment always need a path for the computer program to find a safe way to restart. Automated unloading is often an extension of that approach that brings a calculated closure to the final processing step.

Load-Load Operations

Load-Load Operations is a means of taking significant cost out of the design by having the operators physically travel from one part of the operation to the next. It implies manually taking the part from one machine and loading it into the next. Much of the cost in automated systems comes from the "hand-offs" or transitions from module to module. These transitions typically add no value to the product and are expensive, requiring exact positioning and guidance to be done perfectly every time, at the risk of downtime or damaged parts.

The added benefit to Load-Load is that it engages the operator and often can serve a secondary function, such as a visual quality check. Human inspection is only said to be 80% reliable, but this may be all that is needed for many operations. Some companies have taken this analysis to an even higher level by calculating the return on investment or net present value for the transitions, comparing Load-Load to fully automated transitions. If an operator is already present in the process, engaging them in a transfer and quality check might be the best possible option.

When applying Load-Load you should also consider flow between the steps being connected to minimize buffer inventories. This is a major factor in line or cell layouts. Ergonomic analysis might also be warranted to look at repetitive motion and the range of movement required. Human beings also require breaks and time away from the operation, so back-up coverage will be necessary if the operation plans to continue running without stoppages.

Low-Cost Automation

Low-cost automation focuses on applying lower-cost solutions wherever possible. Many businesses are very cost conscious or cash constrained and finding low-cost means of manufacturing is essential to enable a new product launch. An emphasis on low-cost automation looks for ways of reducing manual interaction with the product or process by selectively adding elements of modest automation wherever possible.

Examples of low-cost automation include use of gravity conveyors versus powered conveyors, or use of lighter-weight, manually assembled supports versus heavy, welded supports. Using simple mirrors to aid inspection of hard-to-reach places, versus camera vision systems is another example. Relays and switches might be used versus programmable logic controller (PLC) driven controls for simple sequences. Where gauging is needed, mechanical or visual checks versus electronic sensors could be used. If a higher level of gauging is needed, perhaps simpler sensors can be used versus more expensive. A through-beam sensor is less expensive than ultrasonic, which is less expensive than a laser, which is less expensive than a vision system. Even with vision systems there are a multitude of options depending on the need for resolution and color identification. Sometimes engineers will gravitate to the latest full-color, highest-resolution camera, when a simple barcode reader could provide 90%, or perhaps all of the requirement.

Material selection can also be challenged. Process engineers and maintenance personnel tend toward expensive construction materials and parts that will last forever, but in many applications a lower-cost alternative is equally viable and can be changed out during routine preventative maintenance periods on a proactive basis. There are endless possibilities to consider while enjoying some benefits of automation at a fraction of the typical automation costs. The only possible limitation is creativity and the lack of ideas.

Mistake-Proof (Poka Yoke)

"Mistake-proof" is sometimes referred to by the Japanese term *Poka Yoke*. It represents designs that make it nearly impossible for bad quality to be produced. In Operations where people are touching the product or process equipment, Poka Yoke helps them consistently perform the task correctly, without the possibility of error.

Use of specific wiring harnesses that are sized to only go where they should, fittings that can only be used to connect the flexible hose to the correct pipeline, or the use of fixturing to ensure parts can only be loaded in the correct orientation are all examples of mistake-proofing. Physical checks such as notching or pinning can sometimes be build into the product or operation. Sometimes mistake-proofing can be accomplished by segregating products to different operations. Mechanical solutions and sensor technologies or barcodes can all be used to prevent errors during manufacture, ensuring quality. Sensors and barcodes are usually associated with gauging and verifying that a step has been performed correctly, but they can be used as part of a Poka Yoke mechanism triggering a gating system to allow the right part or material to be loaded.

Every design must consider the concept of Poka Yoke. The team must determine if it will be critical enough to include as a criteria for evaluating their designs throughout the event, or a less important consideration.

Minimal Capital

Minimizing capital investment is often a goal for new operations and may be crucial to the viability of the product. This is especially true for new product launches where the market acceptance is uncertain. The trade-off between higher initial capital expenses with lower ongoing labor costs may not be an available option. The team's creativity can often discover solutions that find the best of both worlds when cost becomes a major focus.

Several of the other design criteria drive for lower-cost operations. Is there existing equipment that could be shared instead of buying new, dedicated equipment? Perhaps working an alternative shift would provide access to equipment that is too expensive to duplicate for the new line. In situations where product demand is ramping up, making equipment modular might reduce overall capital by enabling inexpensive expansion compared to highly integrated systems.

Often vendors will rent or lease equipment. Consider the financial ramifications of this choice carefully, as it will affect ongoing costs. Your financial organization may also consider this option a "capital lease," depending on the terms and conditions, and it may not create any advantage for you.

The used equipment market can also be an opportunity, recognizing there might be some compromises and risk to evaluate. Applying manual labor instead of automation shifts the costs from initial to ongoing and gives you more flexibility too; however, this is just a trade-off, shifting current costs to future ongoing costs. It could still be the right decision. The overarching goal of this criterion is "creativity over capital," finding new and better ways of performing the necessary value-adding functions at a lower cost.

Minimal Space Required

Minimizing space requirements evaluates designs based on compactness and can be very important if space is limited. In any operation space will be a consideration to minimize travel distances and motion waste as well as costs. Being restricted by a small available space, however, may force the 3P team to use vertical space, often with significant negative cost trade-offs. Vertical space solutions tend to be much more expensive than horizontal solutions due to the cost of the equipment.

As an example, optimizing a distribution center with installation of narrow aisle racking has a very high initial cost, requiring specialized lift trucks or other retrieval systems to operate in the restricted space. There is often a second penalty associated with this type of storage. Tracking inventory in this type of configuration usually requires a highly automated and expensive system because people cannot easily see the inventory to conduct cycle counts for verification. These centers also tend to be extremely slow compared to the pace typically seen in centers with conventional storage and lift truck operators. A thorough evaluation of the trade-offs must be done to ensure understanding of both the positive and negative impacts of going vertical.

Very often 3P naturally creates solutions with less space required without a specific focus on it. Increasing flow between sections of an operation often decreases distances or eliminates storage space for buffer inventories. Right-sizing equipment to meet only the required takt time tends to promote smaller footprints. Correct application of manual intervention, as with Load-Load operations, usually takes less space than automated transitions. All of these elements help to minimize the actual space required.

Low-Motion Waste

Low-Motion Waste considers the flow of the process and applies to movement within the work cell. It applies to the location of tools, materials, control panels, and inspection points that an operator has to interact with. Low-motion waste seeks to optimize both the movement of product and the people who are operating the production process. Incorporating a strong emphasis on process flow generally shortens the product path. Considering the human interaction with the operating line requires analysis of the standard work. Where will a person stand? What will the person be looking at to inspect the product or process parameters? What supplies or tools will the person require, and at what frequency? Any tools or materials that are needed frequently should be kept within arm's length. Items only needed every hour or so might be kept within a few steps. For materials that are needed once or twice per shift, longer distances are acceptable. Supermarkets for resupply of common materials can be set up. Critical gauges or flat screens used for inspection and process control should be mounted where they can easily be seen with as little movement as possible.

When applied at the highest levels, application of low-motion waste principles analyzes the specific movements required for routine tasks. This is done to optimize people positioning and to minimize hand and arm motion, such as stretching, twisting, and reaching. Improving the ergonomics of any given task lowers the risk of repetitive motion injury and improves the quality of the work.

100% Gauging

100% Gauging means building in quality assurance checks following each processing step to ensure that quality problems are detected immediately. Gauging often involves sensors, vision systems, and other means of instantly checking and ensuring that quality product is being made. Mechanical verification can be done too, with strain gauges, micrometers, torque meters, and many other types of analytical tests to assure quality. Gauging is done during production versus waiting until a final inspection to detect a problem.

Gauging can be done using operator inspection though it tends to be accomplished with technological solutions when a high degree of certainty for quality is required. One process might have 100% gauging using operator inspections, understanding that human inspection is less reliable. Another operation might have 100% gauging but use vision systems and machine intelligence, depending on the need.

The use of vision systems and machine intelligence continues to gain momentum, and the technology continues to improve. When applying vision systems it is critical to characterize normal variation in the process and materials being inspected. Vision systems tend to be absolute compared to human inspection that is capable of interpretation. Setting up visual inspection may result in higher levels of rejected "good" product until the system is fully characterized. Operations people who have applied this very powerful technology almost uniformly cite some variation of this story. Other sensor technologies can have similar problems but tend to be more focused and easier to dial in.

How good is good enough? This difficult question must be considered before deciding on any gauging solution. Gauging systems are often necessary, but by definition are not value-adding functions. They add cost and complexity. Though they might inspect 100% of the pieces, they do not work perfectly 100% of the time due to malfunctions or calibration issues. How good *is* good enough? Understanding and applying statistical process control (SPC) to the design of the value-adding functional step may help you improve the capability of the operation to the point where the defect rate is low enough to be acceptable. Would 1 defect per 100 be okay? How about 1 in 1,000? If we could reduce the defect rate to 1 in 1,000,000 would that be good enough? Though gauging is often a very necessary process step to ensure quality, the very best applications of Lean are able to avoid it by building the quality into the value-adding functional step so that it is impossible for a failure to occur.

Maximum Operator Value-Add

Maximum operator value-add looks to preferentially utilize the operator's time on value-adding processes versus non-value-adding steps. Travel time to get parts or materials, conduct inspections, or write up simple reports may be eliminated or reduced if they are considered during design. Some of these steps might be necessary, but they do not generate any value by themselves. Many operations use computer-based systems that store data about the production run. Incorporating an automated report could save time and avoid transcription errors. Setting up kanbans for key materials and locating tools in the right locations avoids hunting and searching during a production run. A focus on this area improves job quality for operating staff and eliminates waste if thoughtfully considered. Maximum operator value-add also reduces stress and frustration, therefore improving morale.

Operator value-add should also be considered in highly automated operations. When automation is applied it often takes away operator responsibility, and when relieved of responsibility most people have less ownership of the operation and the tasks they perform. This is not an insignificant consequence because it will have unintended ramifications. Material variation might cause automated systems to begin rejecting product that an engaged operator would recognize and correct. Consider sources of variation and the power of focused operators, using their senses and ability to process complex information, allowing them to anticipate and correct problems before they happen. Determining ahead of time where those points are, and building interfaces and expectations in, keeps people actively engaged and ensures the other benefits that only thinking, caring people can provide.

Changeover Time

Changeover time is the time lost changing over a line from one product to the next. It could represent a simple variation on packaging put-up or just a lot number change. The frequency of changeover and time lost dictates how much emphasis to place on it. In every operation that requires changeovers, it should be a factor that is at least recognized and accounted for. In operations where product changes will happen frequently, 3P design should address it and find ways of minimizing the impact on capacity and capital cost. Some operations take almost as much time to change over as they do to run the product; they run for 8 hours and then clean and set up for 8 hours. It is easy to see the impact of this. If the operation is at full capacity, they must have double the equipment to meet demand and double the labor. If they have lots of equipment capacity, they will only be doubling the labor cost of the product. Other operations become dedicated to a given product to avoid the lost capacity at the expense of additional cost of equipment. Changeover often has a huge impact on cost and capacity.

Quick changeover is a common improvement strategy that Lean companies utilize to reduce lost time between manufacturing runs and reduce the need for larger batch sizes. Single minute exchange of die (SMED) analysis is another way people refer to quick changeover because much of the technique for this process came from Shigeo Shingo who developed this methodology while working for Toyota. Depending on the frequency and extent of changeovers required in the operation, focusing on this could be very important.

The typical quick changeover event generally involves videotaping the existing process steps to analyze and optimize activities that must happen during the changeover downtime. It looks for tasks that could be done before start of the changeover and those that could be done afterward once the line has started up on the next lot. Much of the improvement generally comes from standardizing the operator or mechanics work and optimizing it. NASCAR pit crews are the ultimate changeover experts, taking the ordinary tasks of changing tires and filling a gas tank to a level of perfection and optimization that few processes ever approach. During a 3P event operators can simulate the steps they will take with the real operation, estimating times and determining locations for key tools or changeover parts to optimize the process. Pinning removable parts, and using easy-disconnect fittings or clamps versus traditional bolting mechanisms greatly speeds up the changeover process. Sometimes additional staff can be brought in from another operation to help for a short period to assist in labor-intensive steps.

Changeover is not just relegated to cleaning and adjusting equipment. In many operations required paperwork, regulatory requirements, computer entries or transportation of quality samples, and other non-equipment-related tasks tend to happen during changeover times. These necessary functions create waste in the form of lost production and should be addressed and mitigated. As mentioned previously, in operations where multiple products are running on the same equipment, changeover will be a very important criterion due to the impact on "each part, each interval" (EPE) scheduling.

Tool Room Maintenance

Tool room maintenance is a consideration in the parts molding industry where molds must be stored, maintained, cleaned, set up, and prepared for operation. There may be variations of this requirement in other industries, but it is a dominant issue in injection molding and similar businesses. Developing systems to manage key tools such as molds so that their status is visible to everyone helps error-proof and increases uptime. Color-coding systems and designated physical locations for tools by status and availability can help to improve efficiencies.

Tooling Quality or Tooling Cost

Tooling quality and the capability of the tools to perform the job adequately and reliability may be another useful evaluation criterion. Where highly accurate tools or tooling is not mandatory, the alternative might be to develop inexpensive tools that are nearly disposable and can be replaced easily.

Consider the options for selecting a paintbrush. Disposable brushes are cheap and can just be thrown away when the job is done. Good-quality brushes cost significantly more and require extensive cleaning afterward. Casual onlookers looking at the completed painted surface might not tell if a disposable brush or a high-quality brush was used. But professional painters will *always* use the better brush. A high-quality brush leaves a far better surface finish with fewer brush marks. It picks up more paint and distributes it more evenly. The cut of the bristles allows for more surgical application at different angles. The weekend handyman will often pick the disposable because of the low investment cost and the ease of clean-up.

Another concern with tooling quality is the tendency to choose tooling that provides higher precision than required. The impact of this is usually higher initial costs. With the higher precision there is often more ongoing cost as well to calibrate, certify, and otherwise maintain the device over its lifetime. Electronic balances or scales are good examples. Higher precision with more significant digits is often perceived as better. However, with higher accuracy comes a need for more stringent calibrations. Isolation from vibration becomes more critical. The installation and ongoing costs will be much higher. Sensors, vision systems, and most tools that can be purchased come in varying grades with many trade-offs that should be considered. Where practical, a statistical analysis should be considered to choose the tool precision required to meet the necessary specifications with a reasonably high process capability. Process capability is described in more detail in the pages that follow.

Tooling quality and cost can be much more than a comparison of trade-offs between similar tools. Innovation and invention can be liberally applied when considering the form, fit, and function of the desired process step. As a final example of how this design focus can be applied, one manufacturer who made polyurethane inserts for shoes captured significant market share by radically changing his model for tooling. They developed the technology to move away from traditional metal molds to a plastic mold that could be made quickly and at a fraction of the cost. As styles changed they could respond very quickly and with incremental investment. Profits rose with higher margins and increased market share.

Safety, Ergonomics, and Health (and Possibly Security)

Safety, ergonomics, and health considerations must be included in the design of every new operation and project. Companies have formal processes to ensure that people are not exposed to harmful materials or equipment dangers. Government occupational safety regulations mandate guarding around moving equipment, restricted access to enclosed spaces, and electrical cabinets.

Safety and ergonomics are also important to every project from a personnel perspective. Some operations will have more inherent dangers than others due to the materials used, temperatures, radiation, or any number of things that could hurt people. A good understanding of these dangers can be developed using standard hazard studies, failure mode and effect analysis, or "new and altered" evaluations. Building these considerations into the criteria for judging designs during a 3P event is important.

The nature of the industry might weight the elements of this combined category differently. If desired, this combined category can be split into two or more separate evaluation criterion.

Security is another factor that could be added to this design criterion. Most companies have standards for security that apply at a total facility level; however, if the street value of the materials produced is significant or if the intellectual property intrinsic to the operation is critical, making provision for physical or information technology (IT) security in a project could be necessary. If important enough, security might warrant special treatment as a separate category.

Environmental Impact

In addition to safety and health, a focus on environmental impact is warranted for many products and processes. Environmental packaging and products that are environmentally "friendly" are benefiting from increasing consumer awareness and preference. From the process perspective, environmental releases are strictly regulated in most countries, and beyond that, many companies have corporate initiatives to reduce carbon footprints and to operate "green" facilities. Renewable resources are a necessary focus for our global future.

Environmental considerations can go far beyond meeting required governmental regulations. Minimizing water or electrical consumption, or reducing packaging materials by eliminating or choosing biodegradable options might be a benefit. Waste streams (discussed below) benefit by minimizing hazardous waste in favor of nonhazardous waste and applying recycling options to virtually eliminate waste. This area is getting more deserved attention, but aside from doing the "right thing" for our environment, it often makes economic sense.

Internal Waste Collection

Most processes generate by-products and have material waste. Designing to effectively remove these waste streams automatically from the process might add great value, versus requiring routine manual intervention. Typically dust, chips, or liquid waste can be managed effectively if provided for in the initial design. Too often these considerations are forgotten because they are not considered a value-adding step, and engineers with less operational experience do not appreciate the impact of routinely having to deal with waste. Many times dust collection systems, simple reject shoots, or bins for capturing minor waste streams are low cost and simple to apply. Other times managing a waste stream could be fairly complex.

In designs where the material is either expensive or in cases where it is hazardous, there could be benefit in reclaiming it from rejected packaged parts. It could open up recycling options with a positive economic impact. In some cases there might be quality implications or regulatory restrictions on how this is done. Under certain conditions special permits required for processing hazardous waste might apply. This type of process can become expensive and might even lend itself to a separate processing step or off-line operation if the complexity becomes too great.

The ability to do internal waste collection could also influence the type of equipment selected. Filtration of liquids can be done in scores of ways. Filter presses, leaf filters, bag filters, cartridge filters, centrifuges, and even settling tanks are all capable of removing solids from a liquid stream. Collecting and disposing the solids that are separated out can be very automated and relatively simple. Depending on the selection, the solution might also be very complex or labor intensive.

There are usually environmental benefits to selecting the best way of managing these waste streams at the point source. Segregating waste close to the point of generation and before it becomes commingled offers greater potential for rework or reclamation than downstream efforts with commingled waste streams.

Simple as Possible

Simplicity can benefit any operation. Simple as Possible speaks to a focus on reducing equipment complexity and making operator engagement intuitive. Simple as Possible processes are less complex and generally more visual. This design target may supercede a focus on benefits gained by flexibility or better process flow. The process steps are kept simple. The operator interfaces are straightforward and simple. There is enough time between required operator interfaces to allow tasks to be completed before starting the next step, avoiding multitasking.

In operations with high turnover, or where multiple operators will be performing complex tasks, the potential for error can be high. Training costs and the time spent developing people to operate complex designs can be extensive. If the skills of the labor force are low relative to the requirement of the new operation, it can be very difficult to train large groups of people well enough to perform complex tasks effectively all the time. Normal attrition and turnover that requires finding new people with the required skill sets and then training them extensively has enormous downsides and a negative impact on the organization that extends far beyond actual training costs and overtime incurred. This often translates to restrictive bidding processes that force people to stay in a job they might not like, causing morale issues or a reluctance for others to bid into these more challenging jobs unless there is a significant financial incentive. This can become an issue in union and nonunion shops alike.

If making fine adjustments and constant tuning of critical parameters is necessary, the initial and ongoing investment in operator training is paramount. Developing standard work in complex situations requires great focus and effort, particularly in operations where the parameters are not well characterized. Rewarding designs that enable simple, accurate processing could be a better initial investment to avoid future performance problems and the ongoing investment in maintaining high skill levels. Simplifying the process to avoid the need for constant tuning and adjustment yields ongoing benefits over the life of the operation.

Standard or Off-the-Shelf Equipment

Use of standard equipment in a project can have many benefits. It can lower capital costs by utilizing commercial equipment that is already competitively priced. Standard equipment is generally manufactured by companies that specialize. They produce a limited range of products and to grow their business they must continually improve their products, usually based in part on customer feedback. Through experience, they understand the equipment's failure modes and willingly sell spare parts and provide instructions for both normal operation and repair. Sometimes they offer specialized training or technical consulting. Availability of spare parts from the supplier lowers carrying costs of parts inventories and has the added plus that the vendor will have experience in recommending the common replacements.

Off-the-shelf equipment normally has cost advantages compared to custom designed equipment. Custom designs are potentially less robust than commercially available options and may require a longer debug phase to reach an acceptable performance level. There is also a familiarity with standard equipment that enables operators and maintenance personnel to utilize their previous experience in operating or repairing it.

Sometimes the necessary features of off-the-shelf equipment are not sufficient for the application. Customizing commercial equipment could be the right compromise, giving most of the benefits of a commercial design and the custom design. These modifications could be done by the vendor or taken in house if the capabilities exist.

Process Capability Ratio (Cp)

The process capability ratio is a statistical measure that indicates how consistent a process is by measuring variation in performance. Every process has variation that affects how consistently it produces the desired results. This is true for an overall operation as well as the individual process steps.

Overall performance reflects the sum of the individual steps. This distinction is no small matter in operations with multiple processing steps. Good design demands that we understand the sources and magnitude of variation in order to access the overall impact on performance and make accommodations. Applying statistical analysis of the critical specifications by calculating the process capability ratio quantifies the risk.

Higher process capability is only better to the extent that it is necessary and adds value to the product being produced. If it does not add value, higher capability may just add cost, because higher precision is often attained at greater expense. For a given process parameter there will be a performance range that the process will naturally produce. There will often be a center point or nominal value that results the majority of the time, and some upper and lower limits in variation, beyond which the process will no longer perform acceptably, resulting in scrap or rework.

Higher process capability values, generally expressed in standard deviations symbolized by the Greek letter *sigma*, implies the least amount of variation and a greater ability to hold tolerances consistently. When using this metric as a design criterion, specify the target variable or variables. Process capability can be applied to production rates, quality performance, and virtually any variable in the process. In design of equipment it can be estimated relative to the specific variables that will influence product quality. Figure 7.3 shows the formula for Cp.

Known Process

Using the known process criterion may seem contrary to the spirit of 3P; however, there are times when it is appropriate. Some companies have an excellent existing process, and there are benefits to duplicating key elements of it versus reinventing a proven solution.

Utilizing a known process can have similar benefits as selection of standard equipment does for the operators, mechanics, computer-aided design (CAD) designers, and others who will interface with it. The application might

$$Cp = \frac{USL - LSL}{6\sigma} = \frac{\text{Upper Spec Limit} - \text{Lower Spec Limit}}{6 \times \text{Standard Deviation (of the process)}}$$

Figure 7.3   The process capability calculation formula.

be the right thing if an existing process is performing very well. Reinvention has appeal, but common sense should still prevail. Just because we *can*, does not mean we *should*.

The caution for this parameter is that it may minimize the creative benefits of 3P, allowing people to gravitate too quickly to a tried-and-true process versus driving for a less costly or more effective solution. There are also cases where the known process is preferred because of its technological advantage as in situations where there is a patented process that the company has developed or licensed. Note also, that although the evaluation criteria are generally applied to all aspects of the alternative designs, they do not have to be. Known process may be applied to only one subsystem if appropriate.

Future Challenge

Sometimes it will be important to design a process in preparation for an anticipated change. This could involve an impending governmental regulatory change or even an intracompany change that is in the works. For regulatory changes, there is often a discussion within industry regarding the potential impact of the proposed changes well ahead of actual legislation. While many of the proposals do not ever get passed into legislation, anticipating them in your design may benefit you with lower compliance costs or other competitive advantages associated with the impending change.

Intracompany change can take different forms. Mergers or acquisitions with other plants or businesses could be impacted by the new design. Having a minimal footprint for the new operation might allow the space needed to enable a future expansion with new product lines. Designing a line that is very modular and portable could be a great benefit if relocation is being considered.

Other considerations for future challenge might be a new technology or capability, emerging competition, new patents, or expiring patents. If selected, this criterion will be based on a known impending challenge that bears consideration.

Maintenance Free

Total productive maintenance (TPM) builds reliability, repair, and maintainability into the initial design. Maintenance free is an approach that considers the needs of operators, maintenance personnel, and support engineers. All equipment will wear and eventually fail. Designing in the ability to quickly service equipment or to make it robust enough to greatly extend the time between failures can be valuable considerations. Eventually every 3P event needs to consider maintainability of equipment and a thorough evaluation of lubrication points, access panels, and maintenance ergonomic considerations before the final prototype is complete.

Maintenance free goes beyond that to consider steps that will nearly eliminate maintenance downtime. It would typically apply to specific critical aspects of the design. In an airplane it might apply to the engine or control system, but would not apply to the tray tables. Maintenance free often takes the form of more expensive materials of construction, heartier designs, or even built-in redundancy, so that a second in-line unit can be activated immediately while the primary is switched off-line and repaired. There are good examples of maintenance free in every industry, typically born out of lessons from previous problems. In the chemical industry, hazardous material storage tanks require special relief valves that cannot be safely taken off-line. Special interlocking valves are used to switch over to a second set of relief valves so that the primary ones can be checked and maintained. Other industries utilize distributed control systems or programmable logic controllers (PLCs) to control critical processes. If necessary redundant PLCs are installed with duplicate software and automatic switches that immediately switch over in the case of an emergency or failure.

Structural considerations may not prevent a failure but could dramatically reduce downtime by enabling a faster repair. Adding overhead beams for hoisting out heavy pieces of equipment or a roof hatch or removable side-wall panels to enable a crane lift are very easy to do during the design phase but are nearly impossible later if not considered prior to construction.

Technical Advantage

Encouraging breakthrough innovation that creates a superior product or manufacturing process is an implied goal in every 3P event. The nature of the 3P process should drive innovation and creative thinking. Adding technical advantage to the design objectives places even more focus on utilizing a new technology that could leapfrog a company because of lower costs, higher quality, or process consistency, or a differentiated product with superior value to customers.

A product or process with technical advantage creates a competitive barrier for entry and can be an important strategic advantage. This could be especially important in highly commoditized markets where current offerings are not highly differentiated. In some cases the new ideas will be patentable, creating an even higher barrier for competition.

Autonomation

The term *autonomation* comes from Toyota and refers to an automated machine's ability to detect a quality defect, stop itself, and subsequently alarm and notify an operator, maintenance, or management to respond and to fix the problem on a timely basis. Without autonomation a machine could be down for extended periods or be producing defective product. With this feature built into a design, downtime is minimized, quality is maintained, and ideally the information gathered on the problem will help to diagnosis the root cause and lead to a more permanent means of eliminating the failure mode.

Examples of autonomation are abundant in most automated operating lines. With computer-based control systems, software is used to trigger equipment sequences, and there are typically sensors or other types of gauging applied to measure that the equipment performed correctly. This built-in feedback mechanism can also be used to trigger subsequent actions such as shutting the equipment down safely, and creating a log of the downtime event so that additional follow-up is possible. Alarming can be done using *andon* systems, a Japanese term for "light" or paper lantern that is used in manufacturing to signal an issue for immediate response and corrective action.

Operations that are more manual have less need for autonomation. However, the principles of autonomation can still be applied to any automated steps within the process.

Development Time or In-House Development

Developing the process using internal resources can create savings and consistency in the design and will often save project time, assuming that adequate resources can be dedicated. All new products and processes have some desired timeframe for delivery, either stated or implied. This is often one of the goals stated in the 3P Event Charter. Applying 3P design has numerous benefits, but there is also a downside associated with the added time required to develop new technologies. Risk management must be maintained rigorously to protect the product launch schedule. Evaluating proposed changes, prototyping, and conducting more extensive testing generally take more time than applying a known approach. There are techniques for speeding up that process, but compared to using established processes, new methods always take more effort. The trade-offs between invention and expediency to market can be estimated for different designs and evaluated; and if desired, they can also be built into financial metrics for comparison.

There are a lot of considerations when evaluating the decision to use in-house versus outside design resources. Understanding core competencies and gauging internal resource availability are crucial. Finding the right balance can mean the difference between success and failure. When attempting to use a new technology there will always be a learning curve to climb. Many vendors are set up to help you with your efforts to apply their technology to your operation and are willing to do much of the development for you. This allows you to outsource the aspects of design with the most risk, and enable in-house resources to focus on what they already know how to do without extending the project delivery time.

The impact of development can be estimated during any stage of the 3P process, but for the final prototype it must be evaluated carefully to establish the anticipated requirements for both time and resources. More radical changes will generally require more time and effort. This criterion may be broken into two separate categories if desired: minimal development time and in-house development. Both address the concept development process, but there is a distinction in their focus.

Scalability

With products that are new to the market and ramping up, or highly seasonal with cyclical demand, scalability may be desired. Some designs allow the process to operate efficiently at a fraction of the full capacity, up through multiples of the target rate. This factor can be highly desired but should not be addressed by simply overbuilding at a higher cost and other negative trade-offs.

Scalability might include modular components that can be duplicated as demand increases. An entire line built to do 20% of the third year takt time requirement in a single shift might suffice for a long time by adding overtime and increasing shifts before a second line is needed. The second line could be a duplicate of the first, or another 3P event might be considered to further improve upon the design, with the option of retrofitting the original. This approach minimizes the initial capital outlay and allows Operations to gain experience before committing to the next phase of growth. Should the product have slower growth in demand or not be successful in the market-place, the loss will be lessened.

Flexibility

Flexibility in design allows multiple options for how and when to use the new operation. For new products, flexibility could mean storing the necessary equipment and rolling it out once a month to run for a day, building the required start-up demand before being disassembled and put back into storage. After a few months it might be brought out to run for several days, and ultimately it might operate full time. The ability of a process to be modular or mobile, skin-mounted or on wheels, with plug-in electrical or hose connections, can give great flexibility and enable sharing or easy storage of equipment. It allows manufacturing space to be utilized more effectively.

Sharing equipment from existing operations is another possibility that saves cost and reduces risk when launching new products. This involves scheduling considerations and becomes more difficult as demand grows. It may jeopardize flow and require inventory buffers initially while scheduling time on the shared equipment. These trade-offs could, however, be an optimum choice for new products with an uncertain demand schedule. Most manufacturing systems tend to be built as fixed monoliths that are not readily changed; 3P challenges that way of thinking.

Flexibility can also involve creating an operation that will support multiple products, including the new product that is being developed. The ability to do effective changeovers and accommodate different products on the same operating line could be very desirable, creating multiple advantages. Examples of this include operating plants with older equipment that could represent a technological disadvantage or a significant risk of failure. Similarly, replacing one or more other operations with a new one designed through the 3P process might effectively reduce space requirements and expand the benefits to other product lines. When considering this criterion always evaluate the full impact on the project to be conscious of any scope creep.

Selecting the Evaluation Criteria

We just considered 30 potential criteria that can be incorporated and emphasized within the overall design of the new product and operation. These criteria will also be used for judging and evaluating the different functions during the course of the 3P event, so they should be considered and selected carefully. Once selected, they will favor the design elements that best incorporate these elements and preferentially promote them.

To varying degrees, most of the criteria will be considered during the event; however, ahead of time, the participants must decide which design criteria are most important. Once the team has considered the different evaluation criteria they must select the most important ones for applying to their 3P process. Typically seven to ten are selected, but this is only a guideline. Selecting fewer will result in weighting the selected factors more heavily, and a larger number will weight each one less. Choosing to use more of them will also take more time going through the evaluation stage.

Using a weighting system that will define their relative importance can help to further differentiate the criteria if desired. A simple "must, should, could" rating system is another option that works well and is fairly quick to implement. A checklist listing the 30 evaluation criteria with those categories is included in the Appendix.

During the subsequent phases of the 3P event, the participants will be divided up into different teams to develop design options. Depending on the team's experience with Lean principles and benefits, the subteams may gravitate to criteria that they are more familiar with. Fortunately, not every event participant needs to be an expert Lean thinker, but ensuring that some people with extensive Lean understanding are part of each subteam will help the group learn together and maximize the design benefits. Limiting the evaluation criteria to the most significant ones will help them focus their efforts and ensure that the inevitable trade-offs will reflect the criteria with the most strategic benefit.

Product Planning and Process Evaluation Criteria Recap

- ☑ Product planning
- ☑ Product evaluation
- ☑ Feature differentiation and prioritization
- ☑ Process evaluation criteria
- ☑ Review definitions of evaluation criteria
- ☑ Prioritize and select criteria
- ☑ Develop weighting of criteria if required

Chapter 8

Defining Value-Adding Functions

Product Features and Process Steps versus Value-Adding Functions

The difference in terminology between a product feature or process step and a function can seem to be a matter of semantics because we tend to think in terms of how we normally do things. When applying 3P, the challenge is to expand our thinking by considering the actual *value-adding function* being performed in the product design, and at every step as a product is made. This is done to avoid biasing our thinking and to expand our creativity. We limit ourselves by referring to a product attribute or processing step in terms of how it is normally accomplished. Jumping to a standard approach to performing a *process step* immediately limits us to one or two possible solutions, not the seven or more that we want to consider.

When 3P is used to develop a product and a process concurrently, there will likely be a preliminary prototype of the product to start with. Ideally there might be numerous examples for the participants to touch and work with. All of the components required to make the product should be available including packaging materials. In manufacturing large-scale products it may not always be practical to have actual parts. The components might be too large or hazardous to bring into an event. Creating a list of all of the materials or packaging containers and adding a picture or a sketch can substitute. Given the opportunity, the participants will find creative solutions to help the group understand the nature of the components when the actual components cannot be made available.

Product Considerations

Product samples, components, voice-of-the-customer information, and accumulated product information are inputs to Lean 3P design that typically come from the overall Lean Product Development process. Consider the product features that create customer value. What attributes are simply there because that is "how we do it" here? Does the color make a difference? Should it ever have to be cleaned or disassembled by the customer? What features will be used all the time, and which ones are superfluous? What specific benefits do the various elements provide? Defining the attributes that are adding value forces us to look at the product solution differently. It opens up our thinking and our possibilities.

A great example of defining product value-add comes from a company that produced clutch and brake assemblies. Their in-house machine shop was very capable and tended to make everything, sometimes simply "because we could." Cost pressures and competition forced them to consider alternate approaches. There was increasing pressure to "offshore" manufacturing to benefit from lower-cost labor rates, making the same product, the same way, in a different place. A cross-functional team formed and started looking at what product features *really* added value and questioned the *purpose* of each feature. They saw that the expensive "housing" was simply *protection* for the coil. It did not need to be "hogged" out of bar-stock at a cost of $3.75 per unit. It could be made of powered metal for pennies. Multiple screws inserted into drilled and tapped holes were used to assemble metal and plastic components, when the purpose was to *fasten*. With no cost increase, protrusions could be provided and ultrasonic welding and rivets could replace screws. The team also reduced the number of major components from five to two. When all was complete, they had reduced the cost per unit from $13.50 to $3.50, making it in the same factory, with the same people.

Process Considerations

Defining value-adding functions also applies to the manufacturing process. To exemplify this point, let's consider the following examples. Applying a label is a process step, but the *function* is identification. Applying a label starts us thinking about printers and types of label stock, while identification opens up multitudes of options. Drilling a hole is a process step; however, removing material is a function. Creating a hole in something with a drill immediately has us thinking about the types of drill bits we might use, or the type of motor that might be needed. Removing material can be done a lot of different ways. We could use a saw, or a grinder, or hydropressure or a laser, or a punch to make a hole, and any number of other means. Process steps versus value-adding functions seem like subtle differences, but getting them right is necessary to remove the tendency to duplicate that which we know. Even if we eventually return to doing a step in a manner that is common to us, making the distinction has merit.

In every plant and in every industry there are standard approaches to doing things. We use lift trucks or overhead cranes to move things. We use injection molding to shape and create parts. We use CNC (computer numerical control) machines to cut and manipulate metal, and extruders to form, robots to place, rivets to fasten, throw-away corrugated boxes to ship. It is easy to forget the basic function that is being done because we automatically leap to the "how to do it" phase. This is the same tendency we face when identifying any problem. People want to jump to a solution, often before we really understand the true nature of the problem. Defining value-adding functions is just the same and can be challenging when an entire group of experts *knows* how it has to be done.

From the chemical process industry another example comes to mind. In one plant there was a large vessel used for processing product after it had been through the key reaction step. The vessel was called a "Deodorizer" and common thinking was that it was there to improve the odor of the product by using a combination of steam and vacuum to pull out the bad-smelling components. This might have been the original intent, but as raw material quality had improved over the years, the need for reducing odor had really gone away. The real benefit of this step today is the removal of other reaction by-products that are not desired in the final product, making this more of a purification step. Once that is understood it opens up the possibility of coming up with other means of purifying the product and removing the undesirable by-products.

3P is a disciplined yet highly creative process. Understanding and articulating the real value-adding functions is a powerful enabler that must be done well because it is the foundation for everything else that comes next.

Evaluating Product and Defining the Value-Added Process Steps

Concurrent Product and Process Evaluation

When possible, organize on a tabletop the parts or sketches in the intended sequence to be used for manufacturing the final product. Component parts can be laid out on either a tabletop or on the floor depending on the size. Pictures, sketches, or scale models can be set in place in lieu of an actual part. The team needs to see and visualize the evolution of the final product from start to finish. Describe each functional step in terms of action words that describe the nature of the function. Do not include transportation or transition steps at this juncture, only value-adding steps. Action words depict what is transpiring in the value-adding step and are written below the parts to describe what is occurring.

The terms used to label the function should be action words. Examples include *attach, bend, separate, make* (hole), *identify, remove, smooth, mix, stretch, guide,* and *hold*. Avoid the use of industrial terminology like *rivet* or *weld, brake bend, centrifuge, drill, label, pick* and *place, index, fixture,* or *pin*. These terms impose a solution or an established manufacturing technique on the function being done. It immediately starts closing down alternative possibility thinking.

This part of the 3P event is where many discoveries and revelations take place that impact design of the product. Even when a product is established and a 2P event is being done to support more demand, this stage often challenges how the product functions or is packaged. In those cases there might be resistance to changing the nature of the product in the marketplace, but changes in packaging may be acceptable. Perhaps the outer carton used to protect the product while in transit or on the shelf could be replaced with a clear blister pack. Maybe there is a more environmentally friendly way of displaying or shipping the unit that would increase consumer appeal or support a corporate initiative. The size of the package might be larger or weigh more than needed. Could you reduce it and save storage space or reduce shipping costs?

When product is being codesigned with process the maximum flexibility is available to make change, but in every case stating the transformational function with action words will inevitably lead to the question "why?" As the team questions and debates the nature of the transformation, new ideas will surface that can lead to profound improvements to the product. This is a key phase of the 3P event because it opens up the creative thinking process and enables breakthrough ideas to be generated.

As new product ideas are generated, the team may be able to mock them up in prototype form to further the evaluation. The right people may be available as participants or Judges in the event, to make a decision, or perhaps it is possible to get customer feedback on the proposed change. You need to evaluate this on a case-by-case basis depending on the nature of the proposed change, risks, and potential impact. At minimum, the alternative proposal needs to be captured on an event "bucket list" to ensure that it will be thoroughly investigated later, and not lost.

Spend the time needed to work this process and if necessary bring in the Judges or other outside experts to vet the new concepts and ideas if it can be done quickly. If it will take too long simply "park" the idea on a flip chart where it can be addressed at a later time. This enables the team to continue working and assure that the new idea is properly considered and brought to a decision point.

With the value-adding steps clearly defined and stated as functions, a very simple flow chart can be created for everyone to see and reference. This can be done on the tabletop or drawn on a wall chart as long as it is visible to everyone involved.

Product Features and Process Step Value-Adding Functions Recap

- ☑ Review product samples, components, and information
- ☑ Organize products to make visible (and physical)
    - ☑ Place the product components in sequential order
- ☑ Describe value-adding functions using action words

3P Example—Kathryn's Finest

Sometimes it is easier to assimilate new concepts using an example. Let's consider a fictitious company as they walk through the steps of the 3P process, highlighting some key points.

Kathryn's Finest

Kathryn's Finest is a large company that makes specialty desserts for the global marketplace. They have several manufacturing plants around the world catering to regional sales. Their products sell in major supermarkets and in convenience stores, and they have a reputation for very high-quality dessert products. They are also known to be innovators in their field, often introducing local dessert concepts to other regions of the world.

Kathryn's Finest is developing a new product offering—a microwavable baked apple. Customer surveys indicated that there is a great demand for this type of product, although there are indications that the interest might be seasonal. Baked apples are popular in all regions of the world where apples are grown. However, on the whole, they are more popular in the United States than in other parts of the world. Kathryn's Finest has done market research and progressed the product concept through the new product development phase-gates, and it has now reached the Business Case phase. They are preparing to invest in a new production operation that will be able to manufacture it.

The company had been using Lean techniques for improving many of their processes and has an interest in applying Lean 3P design to the development of both this new product and the operation that will produce it. They have heard that there are great benefits to using the 3P process. Privately many people are questioning the value that they will gain by going through it and believe it might work in other industries but not in the baked goods business. Others are more optimistic about it, and eventually the momentum has swayed in the direction of going ahead with a 3P event, helped along by the persuasiveness of a Lean-thinking Vice President of Operations, Alexandra.

With the decision made to go forward with a Lean 3P Design event, they contact two seasoned Lean facilitators to help them through the process. A project manager is named and a cross-functional team is assigned to work on the Product Team, chartered with finalizing product design and delivering the new operation. Immediately there is a buzz about the new product and operation, with the engineers and others feeling the pressure to get started. Project team members begin having informal meetings and calling vendors to get budgetary quotes. Others raise concerns about being put on a project team. In the past this has been the responsibility of Engineering and Operations. They already have full-time jobs to do. Why do they have to do Engineering's work too?

Seeing a dilemma starting to develop, the project manager's first decision was to have everyone stop work on all development until the event takes place in about a month's time. Information gathering is still encouraged, provided that no solutions are determined and no detailed design or CAD work is initiated. Some frustration is expressed by the engineering team who must design the operation and by the product development team, who need to develop packaging, presentation, and marketing approaches for the new product. They are already under pressure of a tight timeline and do not have time for a delayed start.

To quiet the growing organizational concerns, a meeting with all of the participants and the people serving as Judges is held to communicate why this new approach had been selected and why it was important to have their cross-functional input. Everyone has a chance to ask questions, challenge the approach, and speak their mind. At the end, everyone understands and is at least neutral to the idea and willing to give it a try.

The Project Manager now meets with the Facilitators to develop the scope of the project and the Event Charter. A copy of their charter is shown in Figure 8.1.

- ☑ Lean 3P Decision Made
- ☑ 3P Team
- ☑ Facilitators
- ☑ 3P Team Communication Meeting
- ☑ Event Charter

THE BAKED APPLE PROJECT 3P EVENT CHARTER

With the charter drafted, a meeting is scheduled with the Process Owners to review it and receive feedback to gain their support. The event coordinators prepare a list of materials to have on hand for building the prototypes. In addition to basic building components and materials, they select some children's books for the event attendees to reference and to help them get into the mindset of 12-year-olds. Although foreign to them, they understand that this methodology is utilized to encourage creativity without limitations and may also serve as an "ice-breaker" to promote teamwork. A work order is created to have someone purchase the materials and have them staged for the start of the event.

The next item they develop is the Event Contract that each attendee will sign on the first day of the event. The purpose of the contract and its importance will be presented in the pre-meeting with the entire team.

The group is fairly well versed in Lean Manufacturing. However, a meeting is set up for many of the participants, to review the key Lean principles. All participants and the people serving as Judges are invited to a Lean 3P orientation meeting, to review the Event Charter and receive an overview of expectations for the week. These meetings will happen in the month prior to the event.

- ▪ Section Summary
    - ☑ 3P Event Charter Approval
    - ☑ Materials Purchased
    - ☑ Nature and "How Things Work" Books
    - ☑ Develop Event Contract
    - ☑ Lean Refresher Training
    - ☑ 3P Event Orientation Meeting

PROCESS NAME
The Baked Apple Project
DATES FOR 3P EVENT
January 16 to January 20
START TIME / END TIME
8:00 A.M. until 5:00 P.M. (working lunch included)
TEAM MEETING LOCATION
Empty warehouse space Bldg 2, and conference room L1
PROCESS BOUNDARIES
Scope of operation begins at Apple Loading and ends at the Shipper Staging Area
Must utilize a space adjacent to the warehouse with given dimensions 90 feet by 52 feet, with a 15 foot ceiling and an 8 foot aisle-way passing through it (can relocate)
Has to be done in 9 months to meet the winter launch
Freezer storage is limited and not desired due to inventory and shelf-life concerns (4 months)
Must adhere to Good Manufacturing Practice protocols
WHY DO WE NEED THIS EVENT?
This product will revolutionize the comfort food market and create a positive revenue stream. Previous product launches were late and ongoing costs higher than expected. We need a better way to design that delivers advantages.
EVENT TARGETS
Cost for new operation must be $1,150,000 maximum to achieve 2 year NPV
Must operate requiring no more than 7 people per shift
Could operate 24 hours per day, 5 days per week, but 16 hours per day preferred
Market studies expect demand to stabilize in year two at the following rates:

	1st Quarter	*2nd Quarter*	*3rd Quarter*	*4th Quarter*
Apples sold	1,350,000	750,000	350,000	1,100,000

Figure 8.1   Lean 3P event charter from the Kathryn's Finest example.

PROCESS OWNERS
*VP Operations—*Alexandra C *New Product Development Manager—*Charles F *Director of Marketing and Sales—*Jon G

PROCESS EXPERTS
*Design Engineers—*Matt D, Philip, Carl J; *Process Support Engineers—*Al R, Bill M, Alan M; *Operations Manager—*Paul G; *Production Supervisor—*Tuan *Operators—*Jimmy B, Stuart; *Material Handler—*Gary P; *Quality Control—*John D *Chief Baker—*Loretta; *Finance—*Rose C; *Receiving/Shipping—*Mike S *Facilities—*Earl; *Procurement—*Nancy Z

EVENT JUDGES
*Chief Operating Officer—*Arvie K *VP of R&D—*Jim L

PROCESS OUTSIDERS
*Farmer Brown Co.—*apple supplier, Jed *Regulatory and Compliance—*Andy

EVENT FACILITATORS
*Lean Consulting Co—*Lisa W *Lean Consulting Co—*Terry M

Figure 8.1 (Continued)

Kathryn's Finest 3P Event—Day One

All of the Process Owners, Process Experts, and Judges gather in the meeting space for the first morning of the event. Introductions are made and after a brief but motivational overview by the VP of Operations, everyone signs the Event Contract. Immediately, the signed contracts are posted on the wall adjacent to the meeting space. The team reviewing the event contract is shown in Figure 8.2. The Team Charter is reviewed, and many questions are raised and sequentially answered. The time limit seems particularly aggressive to the Design Engineers, and because this is the first time attempting a 3P event, there is a lot of nervous banter when the stretch goals are discussed.

When the Facilitators go over the schedule for the week no one is too surprised because they all attended the premeeting and learned about how a 3P event is conducted. Getting down to business, the first step is to

Figure 8.2 Kathryn's Finest 3P team posting contracts.

determine the takt time of the new operation. How many baked apples did they estimate customers would buy once the product is launched? The marketing and sales organization conducted surveys and market research and developed the projections outlined in the Event Charter. Now the raw data needed to be harmonized in conjunction with the capabilities and operational limitations of the manufacturing organization and incorporated into the design criteria of the new production operation.

- ■ Section Summary
  - ☑ Introductions
  - ☑ Signed Contracts
  - ☑ 3P Event Charter Reviewed
  - ☑ 3P Event Schedule

Takt Time

The data from the Event Charter showed that demand was projected to be quite uneven, with stronger demand in the winter months and limited sales in the summer months. During the peak first quarter, the demand for baked apples will be nearly 21,000 apples per day. However, the Operations team has restrictions around working three shifts per day and also recognizes that no operation can run with 100% efficiency. There will be production losses because of quality problems, machine downtime, and for time taken for required meetings.

Table 8.1   Project Takt Time Calculations

Baked Apple Sales per Quarter	1st Quarter 1,350,000	2nd Quarter 750,000	3rd Quarter 350,000	4th Quarter 1,100,000
Option 1				
Work: 5 days/week, 8 hours/day (apples per minute)	72	40	19	59
Calculated takt time (seconds per apple)	0.8	1.5	3.2	1.0
Option 2				
Work: 6 days/week, 8 hours/day (apples per minute)	60	33	16	49
Calculated takt time (seconds per apple)	1.0	1.8	3.9	1.2
Option 3				
Work: 5 days/week, 16 hours/day (apples per minute)	31	17	8	25
Calculated takt time (seconds per apple)	1.9	3.5	7.5	2.4

Takt time has been estimated prior to the event, but they want to let everyone understand the calculation and have a chance to provide new input. Together the team builds Table 8.1 to evaluate takt time based on routine operating schedules and the typical performance of other similar operations. They use the Takt time equation shown in Figure 8.3. They include the following factors:

- *Days of operation*—5 days per week, with weekends possible during peak periods. They tried to avoid Sundays because of a company policy that paid double time for Sunday work.
- *Hours of operation*—Kathryn's Finest ran most operations on a day-shift basis, but a few operations ran two shifts (16 hours).

$$\text{Takt Time} = \frac{\text{Available Time per day (seconds)}}{\text{Average Customer Demand per day pieces}}$$

Figure 8.3   Takt time equation.

■ *Process utility*—On one-shift operations the number of actual productive hours divided by the total 8 hours of the shift averaged 60%. Time lost for start-up and shutdowns, machine stoppages, and weekly staff and safety meetings all contributed to the operational utility. On two shift operations utility rose to 70%.

Table 8.1 indicates the impact of different working schedules on takt time. Although each of the alternatives accommodates the need to produce 21,000 baked apples per day, the hourly takt time significantly changes depending on the schedule they work.

Although the takt time was a fixed requirement, the team recognized that they had options for choosing the working schedule and that the schedule decision would influence their design. They understood that the takt time choice they made would be evaluated along with other criteria, and that the goal was to develop designs that produced at the required rate, and gave lower initial costs *and* ongoing costs, without jeopardizing product quality.

Eventually they chose the 5 days per week, 16 hours per day option, which created the highest overall takt time. They recognized that this would benefit them by improving the overall process utility. It would also reduce the capital costs and ongoing maintenance costs, versus lower takt time options, designed to go much faster, and with increased complexity.

■ Section Summary

☑ Takt Time Calculation

Product Evaluation

The Facilitators wheeled out a cart with apples and all of the possible ingredients for making and packaging the product. The team assembled around a couple of large tables, and they were instructed to start making samples of the product. For the next hour they washed and sorted different types and sizes of apples. They already had a predescribed ratio of brown sugar to butter, but the quantity and size of the hole that they stuffed it in could vary. Cooking temperatures were suggested and wrapping in aluminum foil was optional. For various reasons people hoped to avoid that step in the actual design. There were several possible packaging options that were still being considered ranging from clear-packs to traditional boxes and multipacks. They began trying some different methods of identification. Some packaging was clearly better for protecting the product while in the refrigerated storage conditions. Others simply looked nicer.

The discussions continued to develop as they manually produced the product. This provided interesting new insights, compared to the previous R&D sessions. The Operations people immediately had concerns about some of the initial proposals. R&D was recommending use of different-sized

Figure 8.4   The Kathryn's Finest team developing product prototypes.

apples to vary serving size in the marketplace, as was done with TV dinner products. As this was discussed it became clear to everyone that just dealing with the normal variation in the sizes and shapes of whole apples would be very challenging to manufacture. Immediately they agreed to a standard size and noted on a "bucket list" that they had to set a specification for the standard dimensions for the apples. They would contact one of their process outsiders, Jed, to get his input into what could be achieved through selection at the supplier and get an understanding of potential cost implications.

- ■ Section Summary

    ☑   Product Evaluation

The group also noted that if the hole was too big, it greatly reduced the amount of baked apple the customer would receive and perceived that to be a strong negative. Obviously the entire core had to be removed, but they needed to keep as much of the apple there as possible. Several people were suggesting they add back pieces of apple, in with the sugar and butter, so that the serving size was increased. That idea was greeted with mixed emotions from the others and they decided to place it on the "bucket list."

It was also observed that a baked apple did not really look very appealing with its brown wrinkled skin and distorted shape. All of the functional areas were agreeing that a clear pack could have challenges from a shelf-appeal standpoint; however, they had some experience with that kind of package, and Operations really liked using it. Instead of rejecting the idea

Figure 8.5   The Kathryn's Finest team evaluating packaging options.

outright they decided to allow any packaging concept to be evaluated, recognizing that shelf-appeal was important. Protecting the product while in storage or while in store cold-cases would still need to be considered.

Figure 8.5 shows packaging options the Kathryn's Finest team were evaluating. Figure 8.6 depicts additional product evaluations and feature considerations.

Variations of the traditional baked apple continued to develop. Some people liked adding caramel and putting sprinkles on top. Other people were developing versions with raisins and chocolate chips. Many of these ideas had previously been discussed in R&D meetings and with Sales and Marketing. Now that Operations people were involved they were able to gauge how easily the suggested product variations could be accommodated, at least based on their traditional equipment designs. The extra ingredient features, such as adding caramel syrup, chocolate, or peanuts, all had great interest and the samples tasted great, but eventually the team agreed to reserve those ideas for a future new-product launch once they had gained a better understanding of the acceptance of the original baked apple concept in the marketplace.

They worked through the numerous possible product attributes, with the team capturing the more significant options and rating them by perceived importance and overall manufacturability. Figure 8.7 depicts their decision process.

■ Section Summary

    ☑   Product Attribute Prioritization
    ☑   Product Evaluation Priority Chart

Figure 8.6   The Kathryn's Finest team experimenting with product features.

Evaluation Criteria Selection

With consensus around the major direction for the baked apple product the team now turned their focus to the operating line. However, before starting to consider design options the Facilitators explained to them the importance of first defining the design criteria they valued. They would now develop the Evaluation Criteria that would be used to select the best alternatives during each phase of the 3P process, up to and including the final design. In advance, they would determine their own measures of success.

The Facilitators led the team through the extensive list of possible design criteria, briefly discussing the meaning of each. Many of the criteria were familiar and were currently being employed in other designs they had built. Others were less well understood or completely foreign. On some, they shared different opinions as to the interpretation of what was meant. The Facilitators did a good job of bringing out examples of each criterion and coached the group through the list of design options. Armed with a good understanding of what each criterion meant, the participants began going

Baked Apple Product Evaluation

	Attribute/Feature	Importance Must (3) Should (2) Could (1)	Manufacturing Difficulty Low (3) Medium (2) High (1)	Total
1	Big, Medium, Small Apples	1	1	2
2	Caramel Drizzle	2	2	4
3	Sprinkles	1	2	3
4	Sugar-free Version	2	3	(5)
5	Vacuum Pack	1	2	3
6	Clam Shell Package	3	3	(6)
7	Window Box Package	3	2	(5)

Figure 8.7 The Baked Apple Project–product feature evaluation.

through the evaluation sheet and as a group rated each one. In some cases they spelled out their own definition of what the criteria would be evaluating so that there would be no misunderstandings later.

Rating the different criteria was easier than expected using the "must, should, could" method. There was little debate on the *must* items, but the line between the *should* and *could* sparked some amount of discussion.

POTENTIAL DESIGN EVALUATION CRITERIA CHECKLIST—BAKED APPLE PROJECT

During the rating discussions they jotted down key points defining how they were interpreting each of the criteria, and even developed examples that they felt best described how and why it was important to their design. Finally they decided to include only the *must* and *should* criteria.

Finally they agreed on a list of evaluation criteria as presented in Table 8.2. Figure 8.8 depicts the team during the evaluation selection process.

- ■ Section Summary

   - ☑ Review Process Evaluation Criteria
   - ☑ Rate Process Evaluation Criteria
   - ☑ Determine Top Criteria

3P EXAMPLE—BAKED APPLE PROJECT CRITERIA DETERMINATION
Developing the Value-Adding Steps

Once all of the criteria were established, it was time to identify the value-adding steps of the process. Because they had never made this product,

Table 8.2   Final Evaluation Criteria Selected by Kathryn's Finest

	Evaluation Criteria	*Must*	*Should*	*Could*
1	Takt time	X		
2	Pull system		X	
3	People involvement	X		
4	Automatic unloading		X	
5	Load-load operations		X	
6	Mistake-proof (Poka Yoke)	X		
7	Minimal capital		X	
8	Minimal space required	X		
9	100% gauging	X		
10	Safety, ergonomics, and health	X		
11	Internal waste collection		X	
12	Simple as possible		X	
13	Standard or off-the-shelf equipment		X	
14	Process capability (Cp)		X	
15	Maintenance free		X	
16	Autonomation		X	
17	Scalability	X		

there was a lot of latitude in how to go about manufacturing it, though many in the group had pretty strong opinions of how it should be done.

Going back to the tables with their product prototypes, the group cleaned up and used one table to lay out all of the process steps in a sequence. Within a few minutes they came up with 11 steps. The next challenge was more daunting—coming up with the *functions* being done versus describing the *process* being done. Most of the team was familiar with baking equipment and knew how to do the types of processes that were required. With some coaching from the facilitators, they eventually developed the *functions* that were occurring in each value-adding step.

■ Section Summary

☑   Value-Adding Functions Ordered Sequentially

Figure 8.8 Kathryn's Finest's team selecting evaluation criteria.

Once the participants had developed definitions to describe the functions, they were then able to create a simple flow chart, visually showing the steps required to make their new product. This very simple diagram covered only the value-added steps that were required. At this time, no consideration was given to the transportation or transition steps that linked one value-adding process step to the next.

Table 8.3 shows the value-adding functions the Kathryn's Finest team chose. Figure 8.9 depicts the team physically sequencing the processing steps to make the baked apple product.

■ Section Summary

☑   Sequentially order the value-adding functions
☑   Create a simple Flow Chart

The next chapter covers development of flow charts and begins developing alternative approaches to producing a product. We continue to peek in on our friends at Kathryn's Finest to see how they apply and demonstrate the concepts.

Table 8.3   The Value-Adding Functions

Traditional Process Step	*Description of Function*
Load apples	Accumulate
Sort apples	Separate
Spray apples	Wash (remove contaminants)
Blow off water	Dry
Core apple	Make hole (remove material)
Insert sugar and butter mixture	Fill hole
Bake	Heat
Flash cool	Cool
Insert in clamshell and ultrasonic seal	Protect
Label	Identify
Package (2-packs)	Group

Figure 8.9   Kathryn's Finest physically sequencing the baked apple value-adding steps.

Chapter 9

Flow Diagrams, Developing Seven Alternatives, and Selection of the Better Three Alternatives

Flow Diagram

A simple flow diagram that displays the key functional steps in sequential order will guide the participants through the next few steps of the 3P event process. Review the physical layout of the process used to create the final packaged product. Take each of the value-adding functions and create a process flow diagram on paper. On the whiteboard or paper write the name of the function being performed at each step in the process.

Underneath each step graphically display the function being performed using a simple drawing, a picture, a mock-up, or even by display of the actual parts. Include subassemblies and packaging steps as required, showing how the whole process comes together to create the finished product.

Taking an example from our Kathryn's Finest story, Figure 9.1 represents the type of simple flow diagram required. For more complex processes, or subassemblies coming together, a Fishbone or Ishikawa diagram may be more appropriate to use than a simple flow diagram. The sequence of assembly or fabrication should still be in the correct order, going from raw materials through to packaged finished product. In this case, the steps for

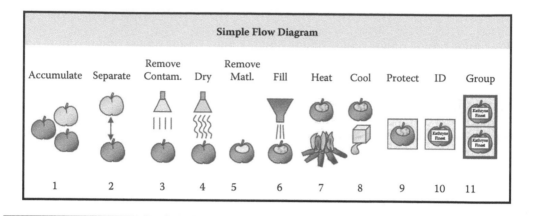

Figure 9.1 Simple flow diagram for the baked apple process.

fabricating each subassembly will be depicted as a branch of the fishbone, and the value-adding steps that contribute to it are listed as offshoots from it. The transformation of the subassemblies via fabrication into the final completed product is depicted showing how the entire assembly moves through each value-adding step.

Some organizations will already have departments or specific areas set up to complete some of the functional steps required. A company could have one area designated with a computer numerical control (CNC) machine used to cut or drill metal parts. They might also have another area with a spray booth for painting the parts, or a location for the parts to be anodized. Another separate area could be utilized for assembly of the parts and yet another could be set up for normal packaging. All of these functional steps might possibly be done in the existing areas, perhaps by adding additional equipment or labor, but it might also be possible to come up with a different design that combines all of the activity into one location—one operation. Seeing the subassemblies helps enable those questions to be asked and considered.

The 3P (Production Preparation Process) challenge is to make the process flow through the steps with as little waste as possible. Consider the ideal scenario. How could you create four smaller lines for a quarter of the total cost, and with four times the productivity over the lifetime of the product? Using a Fishbone diagram the 3P team can see and benchmark the metrics and flow of the "normal" process and see the improvement graphically as they implement design improvements during the event. The example in Figure 9.2 comes from our Kathryn's Finest story and represents a more complex operation shown in a Fishbone diagram.

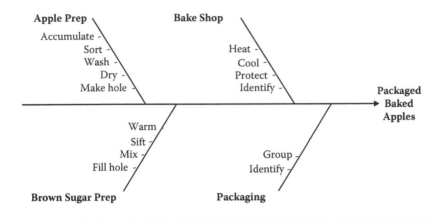

Fishbone Diagram - Baked Apple Project

Figure 9.2   A fishbone diagram can be used to depict complex processes.

There is another possible use of a Fishbone diagram at this point in a 3P event. If the 3P event is being done to add capacity based on an established product and process, a separate Fishbone diagram can be used to capture existing issues and improvement opportunities. This use of a Fishbone becomes a type of "Lessons Learned" document and can help the team avoid repeating existing problems by identifying them early. Listing all safety or ergonomics issues or losses from any of the *Seven Wastes* (waiting, transportation, overprocessing, motion, quality defects, inventory, and over-production) can be very beneficial to the new design. Generally 3P facilitators recommend that a team not go out to the existing operation during an event to look for issues due to the tendency for the participants to be biased by what they see. There is also the risk that it might cause a 3P event to devolve into "just another" Kaizen event. There is a balance here and one alternative would be to have a pre-event team create a list of existing issues with the current operation in the form of a simple checklist or on a separate Fishbone diagram. The 3P team can use that information to ensure that they addressed all of the issues in the new design proposals. Gather baseline data if it is available and if a current product benchmark exists.

Spaghetti Diagram

A Spaghetti diagram is typically a hand-drawn sketch that traces the physical path of materials and people as they perform the steps of a given process. It is an excellent tool for identifying transport or motion wastes in any process. Used in conjunction with the Flow or Fishbone diagrams, the team can look at an actual layout of the space available in conjunction with the plant departments that typically perform that kind of work. With this background the movement of the new product and operation can be traced through those areas, allowing the group to estimate the total travel time and distance using established approaches to completing the functional steps.

In complex processes with many subassemblies or where there is work typically performed by different groups and later combined to create the finished product, a Spaghetti diagram can be very enlightening. Use a layout drawing of the operation and trace the travel of people and parts to show complexity and waste. Calculate this typical travel distance, travel time, and possibly the estimated touch time of the process so that a comparison can be done after the 3P event designs are finalized.

In new manufacturing organizations, where there are no constraints, or in greenfield or brownfield sites, a Spaghetti diagram may not be necessary. When building new operations in existing facility space, it can be very helpful as a benchmark to improve upon. It highlights and quantifies the significant impact of existing layouts of people and overall efficiencies. Figure 9.3 shows a Spaghetti diagram completed by our imaginary friends at Kathryn's Finest.

Developing the Seven Alternatives from Nature

Using the Simple Flow diagram or Fishbone diagram just created, each value-adding function will be displayed sequentially. The 3P event participants now develop seven or more different ways that the function can be performed or achieved in nature. Of all the different elements and attributes of Lean 3P that I have seen in my practice and research, this aspect of 3P has sparked more debate than any other. Some companies see it only as an "icebreaker," while others totally reject it in favor of only considering and comparing industry-proven options. Many companies have embraced it wholeheartedly and find it works very well. Let's consider the benefits.

When we think of natural processes we have to recognize and appreciate the vast complexities of natural design. Nature represents the ultimate in Lean processes. Natural processes only consume what they require, and the by-products, or waste, from one process become the raw material for the next. The exchange of oxygen for carbon dioxide between plants and mammals represents the perfect harmonic cycle. The decomposition of organic

Spaghetti Diagram - Baked Apple Project
(theoretical layout based on current processes)

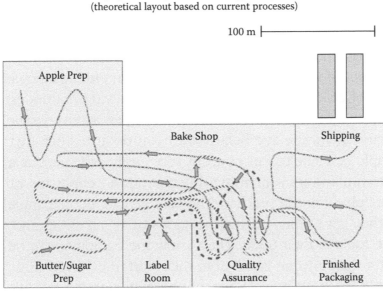

• People-distance traveled: 1,321 m
• Materials-distance traveled: 760 m

Figure 9.3   A spaghetti diagram depicting movement of materials and people.

matter and the eventual formation of oil and petrochemicals, which produce water and carbon dioxide when converted to energy, reveal a simple but powerful transformation process that is representative of the countless natural synergies that surround us.

It seems intuitive that in attempting to develop alternative approaches to solving complex problems, we would consider the engineering solutions right outside our doorstep. To do this effectively we ask participants to think as they did when they were kids. That seems odd and not very business-like, but there are several powerful benefits to encouraging people to develop that mindset. It can be liberating as it enables very serious people the freedom to laugh and think and create without the boundaries that their normal roles often impose. They are asked to think like they did as a 12-year-old, before they learned about industry and equipment and technology.

Twelve-year-olds have been to zoos and parks. The books they have read often educate them in the wonders of nature and how insects and animals behave. They spend time in their backyards watching ants form hills and spiders build webs. They learned about different cultures and climates around the world from the Artic to the Sahara and how people and animals live there. The more ornery ones have learned about solar power while using a magnifying glass to toast an unwary earthworm or how membranes work by applying salt to a slug. Twelve-year-olds do not have much money and to create the impossible things they want, they just take things from their backyard or garage and try to do what they want to. They make pretend toy guns out of sticks and go-carts that roll on old lawnmower wheels. Kids have lots of energy and a passion for fun. They do not understand that the world has limitations, boundaries, and rules. Twelve-year-olds are unflappable and invincible and believe they can do just about anything.

Some people gravitate to this concept very easily and see it as an opportunity to have fun and try something different. Most people are intimidated initially at this phase of the process. They are serious people, intelligent and skilled, and they got to their current positions by being clever and working hard with the tools of their trade. They know the equipment and the technologies better than anyone and before they start, they know how they will end up doing it. Most 3P participants will be willing to try, but the facilitator needs to reinforce the importance of working through the steps and encourage them to see it as the fun, creative process that it is. Continue to remind participants to follow the process and trust that the 3P process will yield the desired results.

Groups that are new to 3P will need some examples of how to apply nature to a function. As an example, if one value-adding step is "attach," you can suggest how things become attached to one another in nature. A fly becomes attached to a sticky spider web, a bee attaches its stinger when angered and the barbs hold it in place. A tick attaches itself to a dog by inserting its cutting mandibles and feeding tube into the dog's skin. If there is a drying step in the process, the example might be of the sun drying out

Biomimicry is a relatively new discipline that studies concepts found in nature and seeks to apply them to real-world problems. Man has always attempted to duplicate natural phenomena. Da Vinci's early drawings of flying machines were based on the structure of birds and bats. We referenced the development of Velcro®, which mimics the burrs on the burdock seed, with their tiny hooks that catch onto fabrics or the hoops in their mating surface. The long and streamlined beak of the kingfisher bird enables it to dive into the water leaving hardly a ripple. It is said to be the inspiration for high-speed trains, helping them overcome air friction to improve energy efficiencies, and reduce noise.

leaves, or a volcano radiating heat from the hot lava. This is the point when the children's books are opened up. Pictures of how things are done by animals and natural processes that happen in different climates of the world can spur different ideas. The children's books also tend to be icebreakers and they allow the participants to laugh at the pictures and stories as they work to think about their seven, or more, alternatives.

The development of ideas from nature should be freewheeling, energetic, and fun. There should be crazy ideas suggested that are tossed around and sketched up. Open-mindedness is the rule. This is the point in the event where crazy, silly, "anything is possible" suggestions are encouraged and applauded. There are no wrong answers and no bad ideas. We are casting a huge net for wild suggestions, and this will be followed by a continual culling of lesser ideas and refining of the best ideas as the 3P event progresses.

Break a large group of all 3P participants into smaller teams with four to six people and divide the functional steps between them so that all functions are allocated. With smaller groups, limit the number of teams and ask them to each develop more alternatives. Each person will first begin to create his or her own list of alternative ideas for the functions assigned to the team. They are all given a pad of larger-sized sticky notes and asked to draw a sketch of the function being performed in nature. They do not have to come up with seven alternatives individually, but should draw all they can think of. In addition to the drawing they should write a short description of the action being depicted. If the function is "shear" and a picture of a crab is drawn, the description might say "claws cut." This step is sometimes intuitive from the picture, but it can clarify the intent, compensating for lesser drawing skills and helping the team group common ideas more quickly. After the individual exercise is complete, all of the sticky notes are put up on the board, beneath the function they represent, and the person who created it explains their picture. The common ones are grouped together. As ideas are generated the participants can place them under the value-adding function on the Simple Process at a Glance, like the one depicted in Figure 9.4.

Encourage the teams to play "catch-ball" with one person suggesting something and another building on it to fully develop the thought. This approach to brainstorming is effective and can be used throughout the 3P process to leverage the individual thoughts of participants and improve them with group input. Often the idea goes back and forth between participants many times, improving with each "catch." The small teams brainstorm until seven or more alternatives are identified. Note that teams often want to give up after only four, five, or six ideas. Do not let them settle for this.

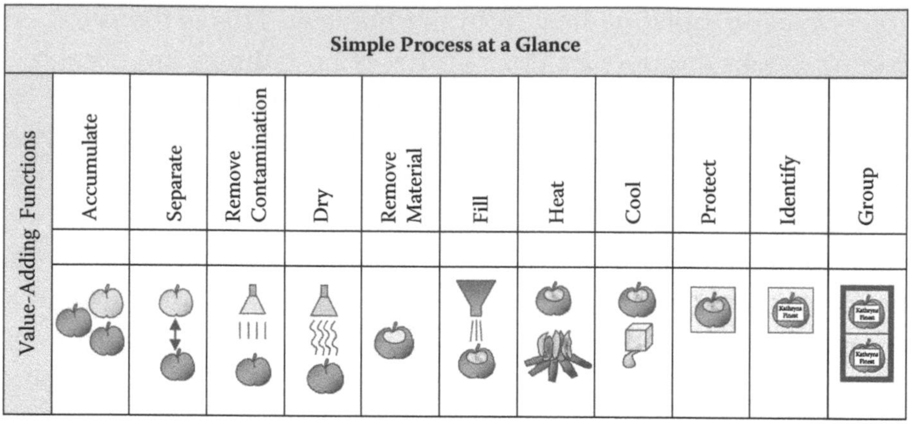

Simple Process at a Glance											
Value-Adding Functions	Accumulate	Separate	Remove Contamination	Dry	Remove Material	Fill	Heat	Cool	Protect	Identify	Group

Figure 9.4 A simple process-at-a-glance guides the development of alternatives.

Encourage them to dig a little deeper or spend additional time researching other methods before giving up.

Before moving on to the next function, each team should ensure that they understand and have noted how the natural phenomenon really works. How does a bee stinger actually get inserted? Why does it stay in place? Add details to every sketch that help explain the mechanism that is happening. The team must fully understand how and why these natural functions occur so that they can successfully develop the idea in the next stage of the 3P process.

Once all of the teams are finished they will present their alternatives to the large group looking for feedback and further upgrades and refinements. Frequently there are functions that repeat during the course of manufacturing the product, and the functions in nature are duplicated. Identification of the product might be an example of this. Often there will be identification directly on the part being produced and additional identification on an outer package. The team might consider developing additional ideas where there is redundancy like this, or should at least consider developing different industrial applications in the next step of the process.

During this freewheeling part of the process, the participants will exhibit a lot of enthusiasm and may joke about the quality of someone's drawing or example. This light-heartedness can be very positive and encouraging but can easily cross the line and become negative. It might be necessary to remind participants not to judge or evaluate the ideas being tossed around. Critical statements or sarcasm will dampen the process and shut people down, limiting the potential results. Usually a reminder of the contract everyone signed is enough to keep the team members on track without impacting the energy levels or excitement this process step generates.

This part of the process will go quickly, often taking 1 to 2 hours to work through and develop the picture ideas from nature. When all of the teams have presented the seven alternatives from nature and developed those ideas with clear understanding of the mechanism involved, it is time to progress those ideas into real-world applications that could work in the intended process.

Considering Industrial Alternatives

Earlier it was noted that exploring natural phenomena as inspiration for our alternatives was not universally accepted. Some individuals and some companies elect to completely abandon nature and go right to industrial alternatives. Here are two thoughts to consider.

If nature is pursued as a first step to developing alternatives, it will certainly open up the group of participants. It is fun and fast and ratchets up the energy level of the group to a very high level. Everyone is engaged. Some of the most profound ideas developed into products have been inspired by nature, and there are new examples of biomimicry being developed all the time. It really works. Having said that, I have also seen a lot of reverse engineering going on during these sessions, with people thinking of industrial processes and trying to relate them back to nature. This is not actually the point of the exercise, but it is okay. Once seven or more alternatives have been developed for each of the value-adding process steps it is also okay to open the gates wider and allow them to consider industrial alternatives. These can come from a variety of sources and different industries. Even if nature is pursued first, allowing industrial options to be added may provide new and different ideas. This is the part of the process where the widest net is being cast. If ideas are not added at this point, there may not be opportunity to introduce them later.

For groups who decide to only use industry as their starting point, the process remains largely the same. A deeper dive into other industries will be done, and from there alternatives will be selected. The success of this approach may lie in the makeup of the team of participants. Eventually the process takes us to the point of evaluating industrially viable alternatives. However, if the team is steeped in their current technologies, they might struggle to get beyond their own niche of experience.

Figure 9.5 revisits our fictitious Kathryn's Finest team, with an example of the seven natural alternatives they developed for their Baked Apple Project.

TRIZ

One technique for helping participants develop industrial alternatives is TRIZ, a process for problem solving and analysis. TRIZ (a Russian acronym referenced as the "Theory of Problem Solving"), is credited to Genrich Altshuller, a Soviet inventor and science fiction author. Astshuller studied thousands and thousands of global patents, across many different fields, and developed a theory that all problems, or contradictions, could essentially be condensed down to 40 "Principles of Innovation." The TRIZ process starts with defining the contradiction. For example, we desire to go fast, but the object is heavy. The contradiction is speed versus weight. The TRIZ Contradiction Matrix is 40 factors wide by 40 factors deep, and the pairing of these elements results in possible principles, or solutions, that have been applied to resolve the stated contradiction. There are many resources available on the Internet that reference TRIZ, as well as several books on the subject by Genrich Altshuller (1973, 1984, 1994) (noted in the Appendix).

Example of Seven Alternatives from Nature—Baked Apple Project

Kathryn's Finest's 3P Team has progressed the seven alternatives from nature for all of their key functions. We look at just one of the value-adding functions: the baking stage, where the product is heated. Figure 9.5 shows what that step could look like.

The first example they came up with relates to how lightning or electricity courses through an item it strikes, turning the item into a giant resistor that resists the flow of electricity and creates heat energy. Their second alternative found in nature represents heat from friction, as when we briskly rub our hands together on a cold day. Molten lava flows over cooler items and heats them via conduction. The fourth alternative uses solar heat radiating from the sun to warm a sunbather. In the fifth alternative an exothermic chemical reaction in a lightning bug generates heat and light. Fire or combustion represents one of the most common methods of heating for their sixth alternative, and the seventh depicts how warm winds heat objects via convection.

In a real-world 3P event the team might consider other alternatives from industrial applications, in addition to examples of natural phenomena. In the case of heat, perhaps they would have suggested induction heating, which generates heat from circulating electrical currents within conductive materials when alternating electrical current is applied to the part itself.

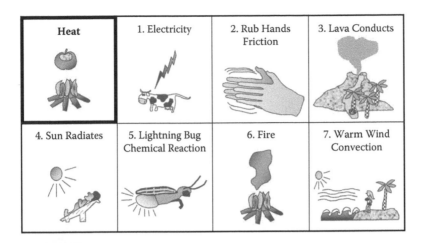

Figure 9.5 Seven alternatives for heating found in nature.

Perhaps they would suggest microwaves, which do exist in nature but are not easily observed. In the latter case we will see how the Kathryn's Finest team incorporates the idea of microwaves into their design, by understanding another related principle—friction.

Kathryn's Finest's 3P Team would have worked through the development of seven natural alternatives for each of the functional steps identified for their process. Once they completed this for all of them, they would be ready to go on to the next part of the process: finding viable industrial applications for each of the selected natural phenomena.

Figure 9.6 shows the Kathryn's Finest team's seven alternatives from nature.

Industrial Application of the Seven Alternatives from Nature

This phase of the 3P process should not be rushed. The quality of the ultimate design hinges on developing these initial ideas and gaining enough understanding of how and why they work to set a direction for the next steps. This is a time to consider what might be possible based on what is done outside of the industry your business is in. Businesses tend to follow other businesses when implementing ideas and concepts such that everyone uses the same approaches for doing similar tasks. Trade shows and industry publications, though beneficial, foster this behavior. Often the best ideas come from a different industry where they had to solve a similar problem

Figure 9.6   Seven alternatives from nature being discussed by the team.

and solved it in a very different way. This type of thinking is what leads to breakthroughs, and it should be nurtured and developed.

With the seven or more alternatives from nature identified, the participants now begin to consider how viable those ideas would be in a manufacturing plant setting. How could this concept from nature be applied to the real world? Going back to the example of "attach" and the pictures of a spider web, a bee stinger, and a tick bite, we can relate to all of those in an industrial application. The spider web stickiness might be accomplished using an adhesive. A bee stinger might be analogous to a screw or a rivet. The tick bite could be replicated using a barbed fitting typical of a laboratory hose connection.

3P participants should work with their small team from the prior step, and again assign each team several functions to develop. The groups need to continue the creative process as they work through each of the alternatives, attempting to find workable industrial approaches. Very often the best new ideas come from completely different industries. In addition, the pace of technological breakthroughs is so rapid that an approach that was impossible several years ago might now be a standard in another application. The power of the Internet search engines to find obscure facts and examples is a phenomenal aid to the creative process.

Teams can search the Internet and call vendors or others within the company, but generally the participants will have enough industry experience that many alternatives will develop quickly. If there are multiple options, they should list them all. Competing companies might make similar equipment applying the same general technology. However, within that scope there might be very different ways that they do it. At this point it is not necessary to understand every nuance regarding the options, but the team should firmly grasp the fundamental basis of the design.

In some cases there will be no good examples of how the natural concept has worked in a factory setting. Do not give up too easily on the research, however. At this point in the process creativity is still the focus, and pushing the team to develop workable ideas might yield new concepts that create strategic advantages. Encourage the teams to play catch-ball with the ideas, developing concepts into plausible solutions. Ask the question "what would it take" in order to make a natural alternative work? This might spur a deeper thought process and lead to a plausible approach.

When the small teams have completed their work developing industrial applications for the natural phenomena applied to each of the assigned functions, they present them to the other participants. The entire group has opportunity to question, probe, and add to the thinking initially developed by the small team. Often new ideas are generated at this stage as new people have a chance to build on others' ideas and many continue to be shaped and refined.

Example of Seven Alternatives with Industrial Application—Baked Apple Project

Looking at the Kathryn's Finest example we see the team has completed developing seven alternatives as found in nature for each of the value-adding functional steps. They are now tasked with taking those concepts and attempting to find an industrial application that uses the technology expressed in each of the alternatives.

Using the previous example for "Heat," we can look at what our Kathryn's Finest team might have come up with using the seven natural alternatives already developed.

1. *Lightning* creates heat by coursing *high-voltage electricity* through the object that it strikes. Because the object being struck is likely to be a less than perfect conductor, it acts like a resistor and heat is generated. As the team researched this, it became apparent that while it is possible to heat food this way, there are no commercial heaters available

that could be purchased, so invention would be necessary if this were chosen. One of the team members suggested that a variation of an arc welder might work, but most of the team remained doubtful.

2. *Rubbing hands together* creates friction. Rubbing apples together does not seem like a very effective way of heating them, but we use microwave ovens every day, and they work on exactly that mechanism. *Microwaves* enter a food and cause the molecules to vibrate, creating friction, which adds heat to the food. The team found many different types of commercially available microwave ovens that could work for this purpose. Some team members were concerned that microwave cooking might not provide the right flavor to the baked apples, but it was still a viable alternative.

3. *Lava conducts heat* as it flows over the objects in its path, and *conduction* is a viable means of heating up an apple. The team considered how deep fryers worked and how the hot oil could quickly heat up food and also imparted a very nice flavor to it. Perhaps they could apply that principle and instead of using an oil, they could use butter, or a combination of butter and sugar. It would change how a baked apple was traditionally made, but all agreed it was interesting enough to consider.

4. *Sun radiates heat* across the solar system, eventually warming the earth and everything it touches. Infrared heaters or other types of radiant heaters and ovens were available commercially. Radiant heat was a viable option.

5. *Lightning bugs* produce both light and heat via a chemical reaction that takes place naturally in their lower abdomen when desiring to attract a mate. Cooking involves many chemical reactions as the organic compounds found in foods change and interact when heated; however, the team could not find any examples where an exothermic chemical reaction could be used for heating their apples in a practical manner. After some research and debate they agreed it was not feasible. The facilitator pushed back and encouraged them to consider an eighth alternative that might be feasible in industry, but time ran out and they elected to eliminate chemical reaction as a viable alternative.

6. *Fire* is one of the earliest heat sources for cooking food. There were many different types of direct fire heaters available for the team to consider which could work with further study, if this alternative was eventually selected. It was noted that costs for this type of heating were relatively inexpensive due to its simple nature, and the overarching popularity of this means of heating.

7. *Warm wind* heating by convection was the seventh alternative, and convection ovens were a very viable and common method of baking. The team investigated several different types of convection ovens including a model that they were already using at the Kathryn's Finest.

With the seven industrial alternatives identified, the team was now ready for the next phase of developing them further before applying their Evaluation Criteria and selecting the better three alternatives they will use to develop prototypes.

- ■ Section Summary
  - ☑ Develop Industrial Applications of the Seven Alternatives from Nature

Developing the Seven Alternatives

Each of the seven industrially-viable alternatives must be developed to ensure a clear understanding. This should be done before judging them against the agreed evaluation criteria. There is some debate among 3P practitioners regarding the decision to develop every one of the functional step alternatives this way, or to first cull the less probable options and only further develop the ones with the greatest potential. A seasoned group is more likely to cull first in order to devote more valuable time to the more promising alternatives. Use caution when making this decision. Every group has some bias toward eliminating the less-familiar technology options. Be sure that the concept has been researched and is understood well enough to make a good decision to drop an alternative, eliminating further consideration.

This is another key juncture for evaluating the *product–process relationship*. Some process alternatives will only provide advantages *if* a specific product feature is incorporated. As you consider the process options, review the product design once again to see if there are additional synergies residing in the optimal selection of process alternatives. Ask yourselves, "if the product was designed *this* way, could we go with a particular alternative that has *these* benefits?"

Right now the goal is to develop the ideas more fully and to gain understanding of each alternative so that a selection of the best three ideas can be identified and progressed to the next phase of the 3P process. This step should be done quickly with the goal of understanding enough to eliminate the least probable alternatives without compromising innovation and creativity.

For each value-adding function you now have six or seven different industrial alternatives. A template for this simple variation of Process at a

Glance is found in the Appendix. Across the top of the first sheet state the title of what that value-adding function is. On the left-hand side list the following, leaving space for a matrix to be developed:

1. *Material* (process step)
2. Sketch of the *Method*
3. *Gauge*
4. *Tool*
5. *Fixture* or Jig
6. *Machine*

The *Title* will state *7 Process Options for* _____. We use an example: Attach circuit board to housing. The function is "attach" but should also make clear what parts are involved and the specific action being taken in this step.

Material describes the process alternative. Continuing the example of "attach," the first alternative might be "adhesive" and a second might be "apply screws."

Method includes the sketch of each alternative across the top row of the table. In the example there might be a sketch of the housing with the circuit board set in place and some sort of adhesive being applied to the edges. In the second case there might be a picture of an automatic screw gun screwing the board down.

Gauge refers to a process check to verify that the step was performed correctly. In this example, perhaps there is a torque measure to verify that circuit board is firmly attached by the adhesive. There might be a sensor to ensure that all of the screws were set in place or a torque sensor in the screw applicator.

Tool describes any devices required to perform the work. If this step is performed manually the tool might be the screw gun or adhesive applicator. Perhaps there is a tool to determine removal torque that will be used.

Fixture or *Jig* focuses on how parts are held accurately in place. The step might require a pinning system or a nest to hold a housing in place. A guide could be needed to ensure the circuit board is positioned in the exact location required.

Machine describes attributes of the equipment used in this process step. It defines if it is simple or highly complex or automated. It might describe guarding needs or attributes needed to maintain it. For the example, the adhesive might use ultraviolet light to cure and require a shield to protect personnel exposure. Similarly if a robot is used to apply the screws a physical barrier or light curtain would be needed for protection. There might be lubrication points desired to maintain the machine easily.

Going through these steps for each alternative and for every value-adding function can be a very tedious process. This is why experienced 3P teams will typically eliminate the clearly inferior alternatives before wasting time to develop them further.

After the three top alternatives have been selected, the teams will continue developing and documenting information on the functional process attributes using a *Process at a Glance* chart. This will be further developed for each of the three prototypes in the next phase of the 3P process.

Example Developing the Seven Alternatives—Baked Apple Project

Kathryn's Finest's 3P event was now in full swing. They had come up with good industrial applications for six of the seven alternatives and developed their knowledge of those functions, documenting the information on the Process at a Glance chart.

They did not have all of the information to fill in every box, but the facilitator reminded them that this was okay. The goal here was to better understand the alternatives and consider underlying benefits or complexities that might help differentiate the better three from the rest.

When they looked at the first alternative, using a type of arc welder to electrically cook the apples, they recognized that protecting people would be a big issue. One of the electrical engineers also noted that they would need to have a special nonconductive basket or holder for the apples to contain the electricity where it needed to be. Several of the other technologies also required protection for people. The concept of using a deep fryer for heating the apples in warm butter and sugar would need a tool of some sort, perhaps a basket, to drop the apples in and then retrieve them when they were cooked.

As the group was researching and discussing the attributes of the seven alternatives, they were developing charts for each value-adding process step. The example shown in Figure 9.7 is one they developed for the heat function.

Now they were ready to learn how to apply the criteria and select the better three alternatives.

■ Section Summary

☑ Develop a Modified Process at a Glance for Each Function

	7 Process Options for: Heat						
Material	High Voltage Electricity	Microwave Heat	Conduction Heat	Radiant Heat	Exothermic Chemical Reaction	Fire	Convection Heat
Method	Arc Welder 	Microwave Oven 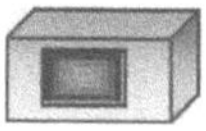	Deep Fryer with Butter 	Infrared Oven 	??? X	Pizza Oven or Broiler 	Convection Oven 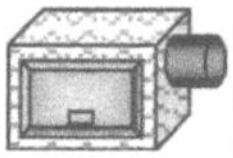
Gauge							
Tool			Basket type device?				
Fixture or Jig	Special non-conductive holder						
Machine	Personnel protection concerns	Personnel protection concerns	Thermal protection			Thermal protection	Thermal protection

Figure 9.7 Kathryn's Finest process-at-a-glance for the heat function.

Figure 9.8 shows the Kathryn's Finest team reviewing the seven Alternatives for "Heat Process at a Glance Chart."

Evaluation and Selection of the Top Three Alternatives

The 3P participants should now have a very good understanding of each of the seven alternatives for each of the value-adding functions. The process for selecting the better three of the seven can be done in various ways but should incorporate the Evaluation Criteria selected in an earlier step. Allow yourself the ability to again eliminate any obvious alternatives that should be cut before going through a thorough evaluation. The trade-off between thoroughness and expediency has to be considered in every event.

This is an appropriate time to bring the event Judges (Critical Evaluators) in to participate *with* the team in narrowing down the choices from seven

Figure 9.8   Kathryn's Finest reviewing a process at a glance chart.

to the final three for each value-adding function. Take time to orient the Judges to the concepts and how the concept development progressed from examples from nature to industrial applications. It is important to gain their alignment with the direction in which the team has progressed before going on. Some course corrections are still possible but are not typical as long as the team has followed the 3P process and developed the facts that support each decision and alternative.

The Pugh Method or Decision Matrix

An effective and very simple method of evaluating alternatives can be done using the Pugh method. Stuart Pugh was an extraordinary design engineer and manager whose work in developing essential design concepts could potentially be the foundation of the Lean 3P process. In his seminal book *Total Design: Integrated Methods for Successful Product Engineering*, which was published in 1990, Pugh states: "Total Design is the systematic activity necessary, from the identification of the market/user need, to the selling of the successful product to satisfy that need—an activity that encompasses product, process, people and organisation."

Though there are variations on the Pugh method, the simplest approach seeks to compare the alternative to a similar standard process if one exists. If there is a similar existing process that can be used for comparison, judge each alternative against that. If the alternative is expected to perform better than the existing process step, mark that specific criteria with a **+**. If it is about the same give it a 0. If the alternative is not expected to be as good as the existing process for that criteria give it a –. Repeat this process for each alternative. Some versions of the Pugh method use "S" for same or similar instead of "0," and others use +1, 0, or –1. Choose a version that works best for your team.

During the evaluation, everyone participates in rating the features and has an equal part in the selection process. The people serving as Judges and other senior managers may have the ability to influence the selection process but endeavor to keep the discussions fact based and encourage involvement by the full team of participants.

In some cases, one evaluation criterion may have a greater importance than others, so a simple weighting factor could be used to give it more emphasis. If it is twice as important you could add a 2× factor and give it two pluses or two minuses depending on the rating. Another option for increasing priority of the criteria is to use a range of one to three pluses and one to three minuses. Some criteria would only be valued at one plus, while others might be valued at three.

When each of the alternatives is evaluated, score the number of pluses and minuses at the bottom of the sheet and look at the concepts with the highest number of pluses. Do they have any minuses? Could those minuses be mitigated somehow? Go through this evaluation and have the team select the highest-rated three alternatives. Repeat this process for each of the value-adding functions.

To exemplify the Pugh method, consider an evaluation of different means of transportation. Listed below are four options—automobile, bicycle, rickshaw, and hot air balloon. Also included on the left-hand side are the evaluation criteria—initial cost, ongoing cost, travel time, and safety. The team rating the four alternatives might have rated them as shown in Table 9.1.

Looking at the alternatives, the automobile is just as positive as the bicycle but not as negative, so it comes out as the top choice. However, a car and a bicycle have completely different costs. If initial costs are more important than the other factors, how does our matrix work out with a weighting factor? Table 9.2 shows how the Pugh method is used with a weighting factor.

Table 9.1   Rating Alternatives using the Pugh Method

Means of Transportation	Automobile	Bicycle	Rickshaw	Hot Air Balloon
Evaluation criteria				
1. Initial cost	–	+	+	–
2. Ongoing cost	0	+	–	–
3. Travel time	+	–	–	0
4. Safety	+	–	–	–
Total +	2	2	1	0
Total –	1	2	2	3
Rating (1 through 4)	1	2	3	4

Now the bicycle is favored over the automobile based purely on the evaluation criteria with a weighting factor. Through this simple example, we can see the impact of criteria evaluation on the selection process. Realistically, a car is pretty different than a bicycle, and other factors might influence the decision. In our real-world examples we may encounter a similar situation, in spite of having a good evaluation process. There will always be other intangibles not specifically addressed in the criteria. We explore that idea while again revisiting our example from Kathryn's Finest.

Table 9.2   Rating Alternatives using the Pugh Method with Weighting Factors

Means of Transportation	Weighted Priority	Automobile	Bicycle	Rickshaw	Hot Air Balloon
Evaluation Criteria					
1. Initial cost	2×	–	++	+	–
2. Ongoing cost	–	0	+	–	–
3. Travel time	–	+	–	–	0
4. Safety	–	+	–	–	–
Total +		2	3	1	0
Total –		1	2	2	3
Rating (1 through 4)		2	1	3	4

Example of Selecting the Better Three Alternatives—Baked Apple Project

The 3P Team at Kathryn's Finest has spent most of the afternoon of their first day developing technologically feasible applications of the seven alternatives and done enough research to understand how they could work. In some cases, like in using high-voltage electricity to cook an apple, the concept could not be duplicated using existing equipment. A couple of the R&D guys had become pretty excited about the concept and had determined that it might be worth exploring, but suspected that electrocuting apples might be lacking in culinary appeal. They still elected to allow the idea to work its way through the evaluation process. The team developed the process for their 11 value-adding functions:

- Accumulate
- Separate
- Wash
- Dry
- Make hole
- Fill
- Heat
- Cool
- Protect
- Identify
- Group

Kathryn's Finest 3P Event Judges arrived for the discussions. Initially the Facilitators and the Process Owners went over the sequential processes that the team had used during the day, describing how the concepts were progressed and answering questions. There was a lot of good-natured kidding as they showed the Judges the various sketches the participants had made to demonstrate their concepts. Though the Judges seemed a bit skeptical of the process, they were clearly impressed with the ingenuity and creativeness of the team, and the excitement and enthusiasm being demonstrated was contagious. None of this was typical of normal business at Kathryn's Finest.

The team got down to business right after the orientation. They approached the first step of the process (Accumulate) and put up a hand-sketched evaluation chart. The chart listed most of the Evaluation Criteria that they had previously agreed were either "Musts" or "Shoulds." They started comparing the proposed alternative with a typical approach they would normally use to accomplish it, if there was one. A few of the lesser "Should" proposals were left off the list at this stage in order to save a little time. They would be used later when evaluating the prototypes. There was considerable discussion around some of the more interesting approaches, but overall the process went relatively quickly.

When the subject of the baking step came up, the evaluation became more animated as this was a critical step that not only broke down the structure of the apple through the cooking process, but it also imparted different textures and flavor elements to the final product. The goal of this step was to partially cook the apple, leaving the final cooking step to the customer when they warmed it up in either their oven or microwave. Even a better taste might not be acceptable if it differed too much from what customers might expect it to taste like. Already that afternoon some preliminary skunk-work was being done in the food-testing laboratory to see what the impact of a few of the technologies might be. Recognizing that the 3P event was likely going to need support, the laboratory had been put on stand-by to support them as needed.

The Judges and many of the participants met the first alternative, using high-voltage electricity to heat the apple, with much skepticism. Still they worked through the criteria, recognizing that even if it was rated highly using the criteria, the food testing taste evaluation would be a major hurdle to clear. This approach would certainly be faster than normal baking, giving them a "+" in the takt time category, meaning that it would not limit the overall takt time of the operation. However, there were questions about how you would error proof such a new technology, so it received a "–" score there. In many categories the team estimated that it could be the same as the traditional process, but in others it also scored lower, including safety, higher capital cost, and ability to apply Jidoka or autonomation. When complete it only rated one "+" and had four "–". Several team members groused that the time spent on this option had been wasted. Others quietly agreed but stayed silent because they were all new to the 3P process. They rationalized that it had only distracted them by a handful of minutes, and there was always a chance that it could have had advantages.

Using a microwave oven approach was not typical but was very possible. After all, this was the approach that the customers would use to finally heat up the product. They knew that microwave cooking was different from other cooking methods and often resulted in less flavor, but the food scientists had ways of enhancing the flavor naturally, and overall it was considered an achievable alternative. This option looked very good in the evaluation also, with positive takt time, ability to error proof, and the space required all getting pluses. The only negative was a potential safety issue due to the microwaves. They would need to add physical shielding to ensure the protection of the people operating the line.

The third alternative was also very innovative. Using a hot butter and sugar mix to "deep fry" an apple seemed crazy, but it was certainly doable. One of the Judges reminded people of the first time someone deep fried a Thanksgiving turkey and how it caught on like wildfire. It was unconventional, but the possibilities seemed very interesting. They continued through the criteria for the alternative and added up the pluses and the minuses. Overall the scores were pretty good.

The remaining three alternatives were evaluated in the same manner. After developing the raw scores the team continued to discuss the possible best options considering how they might mitigate the negatives and overall how positive the pluses really were.

When they completed, they decided on the three better alternatives to be used to heat or cook the apple:

- Deep fry
- Convection (conventional oven)
- Microwave

■ Section Summary

    ☑ Summarize Status for the Judges
    ☑ Use the Pugh Method and Rate the Alternatives
    ☑ Select the Three Better Alternatives

Figure 9.9 shows the Kathryn's Finest team selecting the three better alternatives for "Heat." The better three alternatives for Heat, and the better three from the other 10 value-adding steps would now be processed into actual prototype operations. The participants were anxious to see how the

Figure 9.9 The team completes the evaluation of the 7 alternatives for "heat."

	7 Process Options for: Heat						
Material	High Voltage Electricity	Microwave Heat	Conduction Heat	Radiant Heat	Exothermic Chemical Reaction	Fire	Convection Heat
Method	Arc Welder 	Microwave Oven 	Deep Fryer with Butter 	Infrared Oven 	??? X	Pizza Oven or Broiler 	Convection Oven 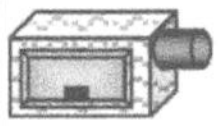
Gauge							
Tool			Basket type device?				
Fixture or Jig	Special non-conductive holder						
Machine	Personnel protection concerns	Personnel protection concerns	Thermal protection			Thermal protection	Thermal protection

Evaluation Criteria	Arc Weld	Microwave	Deep Fry	IR Oven		Broiler	Convection
1 Takt Time	+	+	+	o		+	o
2 People Involvement	o	o	+	+		o	+
3 Poka Yoke	−	+	o	+		−	o
4 Minimal Space	o	+	o	o		o	o
5 100% Gauging	o	o	o	−		o	o
6 Safety, Ergonomics	−	−	o	o		o	o
7 Scalability	o	o	+	o		o	o
8 Pull System	o	o	+	o		o	o
9 Minimal Capital	−	o	−	o		o	o
10 Autonomation	−	−	o	o		o	o
Total +	1	3	4	2		1	1
Total −	4	2	1	1		1	0
Rating (1–7)	6	(3)	(1)	4		5	(2)

Figure 9.10   Evaluation of the 7 alternatives for "heat" using the Pugh method.

functions would be distributed out to the three teams. They were even more curious about to which team they would be assigned.

The Judges left with a clear understanding of the potential benefits of the proposed better alternatives and supported the efforts to progress them further in the next stage of the 3P process. Having their support brought a sigh of relief to the participants, affirming that they now had permission to explore some pretty different alternatives.

Figure 9.10 represents the results of the team's evaluation of the seven alternatives for the Heat process.

Summary of Progress

This point in the Lean 3P process typically comes between the end of the first day and the middle of the second day. Reflecting back on progress so far, we have now developed the scope of the 3P event and grounded all of the participants in the methodology, event goals, and boundary conditions. We developed and selected the evaluation criteria that most fits the project goals. The team then examined the product to be produced and challenged its functionality and features and identified the value-adding functions that transform the raw materials into the finished product.

We first developed seven alternatives based on how these value-adding functions occur in nature, and reflected on the near-infinite possibilities just as we would have when we were 12 years old. We considered the mechanisms in place that made the natural phenomena possible and learned about how they worked. Then the team began converting those ideas and concepts from nature into tangible, technologically feasible approaches. Details of the industrial processes were developed further to gain significant understanding of the attributes and considerations associated with each one. Where possible, additional industry-based alternatives were added to the alternatives based on natural phenomena. Finally the team applied the

A NOTE ABOUT EMOTION

Lean 3P is a high-energy process, exercised with a large group of people moving quickly through a structured series of steps to achieve very big and important goals in very little time. There is an ebb and flow to the process that is difficult to capture, until you experience it. Every event has high highs and some deep lows. People get physically and mentally tired as they balance the joys of unleashing their creativity, with the pressures of hard delivery dates. It happens to some degree in every event, and when it does, we just have to remind ourselves to trust the process. You will work through it, and in short order, the team will be rushing forward together to meet the next challenge. There are no suggestions to help you avoid it or prepare for it, beyond recognition that it is a natural by-product of the 3P process intensity, and the passion that it fosters. It is that same passion that fuels the innovation, creativity, and pursuit of excellence within all of the stakeholders that makes Lean 3P so uniquely successful.

judging criteria used for evaluation and rated each idea. Seven alternatives converged down to the better three process concepts for each of the functional steps.

Now with three robust alternatives selected for each value-adding function, the 3P event can enter the next phase and begin developing the three prototypes.

Flow Diagrams, Developing Seven Alternatives, and Selecting the Better Three Alternatives Recap

☑ Flow diagrams
☑ Spaghetti diagram baseline
☑ Develop seven alternatives for each value-adding function
   ☑ Develop alternatives from nature
   ☑ Convert to industry-viable alternatives (add additional if desired)
   ☑ Evaluate alternatives using agreed Evaluation Criteria
   ☑ Pugh matrix

Chapter 10

Process at a Glance, Selecting Teams, Constructing the Three Prototypes

Process at a Glance

The Process at a Glance is a means of visually depicting the value-adding functions in the sequence used to produce the finished product. It ultimately becomes a living document that is continually refined and updated over the life of the project. The full Process at a Glance incorporates all of the elements used in the preliminary evaluation criteria used to narrow the seven alternatives down to the top three: Material, Method, Gauge, Tool, Fixture, and Machine. These key aspects of design act as a checklist for the design team, ensuring a holistic approach that increases probabilities of success.

Now that three alternatives are available for each of the functions, we create three separate Process at a Glance charts. The alternatives should be reviewed and consideration given to forming combinations of alternatives that might work well together to create a specific advantage. If there are no obvious combinations, then simply order them sequentially. Figure 10.1 shows how the teams will be assigned functional alternatives to develop from the three available choices.

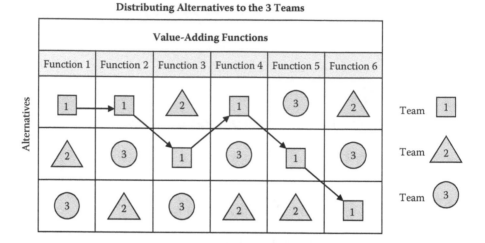

Figure 10.1   The assignment of alternatives to the three teams.

Team Selection

Before the event starts the three teams should be selected. This saves time and also gives opportunity for more thoughtful pairings. There will be stakeholders from many different functional areas and from different levels of the organization, from very tactical jobs such as material handlers and operators to more strategic jobs with managers or directors. In the event, however, all participants have an equal voice and there is no advantage placed on job ranking. Having said that, there will be personality considerations and skill considerations to take into account when putting together the teams charged with building the three prototypes. Take care to evenly disperse expertise and experience on each team.

Engineering capability is a skill that is heavily utilized in this phase of the 3P process. Every industry will require different engineering functions depending on the nature of the business, so there is no guideline for how many to have or what types of engineers are best suited to be on the three teams. Each team should have a good representation of the required skills needed to research the alternatives and develop the verification criteria necessary to endorse or reject a particular functional alternative. Rarely does an alternative get completely rejected at this point in the process. Earlier steps have progressed the understanding and every alternative *should* be viable before beginning to build the models.

With the engineers balanced and allocated to the three teams, the other members will be filled in, again attempting to achieve balance between strategic and tactical functions. Lean thinking and awareness might be another consideration for how the people have been allocated. Overall the goal is to have numerically and functionally equivalent teams working on each prototype. Avoid allowing the teams to trade people or make changes unless there is a very good reason for doing so. Recognize if an individual has deep technical knowledge or a strong passion to explore a particular alternative. Either assign them to that team, or allow the team to which they were assigned to work on that alternative.

The question of team leadership can be kept informal, allowing the natural leaders to evolve during the process. Groups often defer to a senior person for guidance, and there usually seems to be one or two people who have a natural tendency toward organization. The Facilitators need to keep watch over this dynamic and ensure that the teams get out of the gate quickly and also ensure that no one person is dominating the discussions and direction. This sense of balance in group dynamics may work itself out effortlessly; however, this is a critical stage of the process, and it requires attention to make sure *all* ideas are developed. It may be practical to select and name a team lead for each of the teams depending on the overall personalities of the participants or the company culture. In 3P events I have overseen in the past, strong leadership was able to take their teams to highly creative and effective prototype solutions, while others with equally strong leadership went in a sideways direction for too long, wasting time and not creating their complete prototype. This eroded the potential value of the event and the effectiveness of that team for the rest of the 3P event.

Consider the prototyping phase of the Lean 3P event as a small project. There is a significant amount of work to be done in a very short period of time. Research must be continued for each alternative. Vendor visits might benefit the group and will need to be arranged quickly. The physical construction, evaluation, and reconstruction (with improvement) of the prototype will consume a lot of time. A high-level project plan with some intermediate milestones will help keep the teams synchronized and on pace. Typically the team leads and the facilitators will closely monitor progress. Team members will be very cognizant of their deadline too, often working long hours during this phase in order to fully develop their alternatives and ideas. There is an informal competition between teams that drives up the energy levels, and people naturally push themselves very hard to be successful.

Post the team names on flip charts by their empty Process at a Glance sheets in different areas of the overall workspace where they will be assigned to work. The teams will assemble together and start reconstructing their Process at a Glance parameters on their sheets or whiteboard and begin discussing the approach they will take.

One final note on the team selections is that throughout the 3P event there is a conscious attempt to keep switching people to different teams and different alternative ideas. This is done to avoid people's natural desire to take an idea they developed and run it from start to finish. It attempts to minimize the normal biases toward the ideas they came up with, and forces people to consider others' ideas. Switching up of the teams at this part of the process can be valuable to the creative process and ultimate success of the project. Temper this point with some judgment. If a person has a truly brilliant idea, with loads of promise, and a strong passion to pursue it, letting that person run it from start to finish might be exactly the best choice.

Safety Reminder

Before each team rushes off to develop their ideas and begin building their prototypes, it is important to cover appropriate safety precautions. Some participants may have lived their lives and never touched a power tool, razor knife, or hot glue gun. In every event there will be some people who have experience with these tools and how to use them. Establish firm safety rules for using the tools and enforce those minimum standards rigorously. Provide some training or restrict access to certain tools if necessary. Nothing would derail a 3P event more quickly than having one of the participants get hurt, and your people's safety must supersede any other event deliverable.

Common safety guidelines include

- Safety glasses with side shields when using tools
- Steel-toed shoes when lifting or moving heavy objects
- Enforcement of good lifting techniques
- Kevlar® gloves when using razor knives or cutting tools
- Thermal gloves for using hot glue guns
- 5S concepts should be applied to maintain workplace organization with routine clean-up
- Good lighting
- Enough space to set up and operate in an organized manner

It may be appropriate to prepare these safety rules ahead of time and have them posted in the prototype area. Some companies include a version of the safety rules in the contract people sign at the beginning of the event.

Getting Started with the Three Prototypes

With the teams formed in their respective areas, there will likely be questions about what to do next. More mature 3P teams may not require much grounding before getting started. Most teams will benefit by a quick overview of the expectations and a review of the evaluation criteria. They must know the time expectations for when the evaluations are targeted to happen and how far they need to be in development of their prototype when it will be assessed. This step of the 3P process will inevitably have many trade-offs. Choosing carefully when to drive deep into the details and when not to will be an ongoing tension as the teams research the functional steps and start to build the prototypes.

The first actual task for each team is to transfer their assigned process method alternatives over to their Process at a Glance. This is a good opportunity to develop the team's collective understanding of the steps they will be pursuing. During those preliminary discussions a good sense of the relative difficulty in each step should come out. Is this a new technology? Will the scale of this operation make other applications of this concept relevant? How could we make it work? Who are the subject matter experts on this technology that we can talk to? All of these questions and more will surface during the preliminary discussions.

Utilizing the Assigned Space

Each of the workplaces should already be marked off with the natural boundaries of the targeted physical space for the new operation, assuming one exists for the process being built. Physical constraints such as walls, support beams and columns, stairways, overhead limitations, doorways, and fixed aisles should be taped off, along with any "immovable" pieces of equipment or other monuments. Do not be afraid to challenge some of these barriers as most things can be changed with enough time and money, but avoid creating a space that will prove to be problematic due to unrecognized constraints. It is better to include the constraints and fight to justify removal of them than to miss considering them at this stage. Again, these constraints should be identified before the event start whenever possible.

In cases where there is no defined location or in brownfield or greenfield sites, there may not be *any* physical constraints identified; however, recognize that every facility has constraints due to support columns, utilities, and a range of other considerations. Consider having the new facility design team part of the 3P event and allow them to relate your ideas to theirs. If the process being developed is physically large enough to preclude building full-sized prototypes (or if it is simply not practical to develop full-sized prototypes), it is possible the teams will be building small-scale models and need only a large tabletop set up. A team may decide to build multiple smaller operations (for lower cost) to meet demand. In those situations they could build one prototype using a fraction of the allocated space and explain that the remaining space is for repeated models.

The Need for Speed

There are no rules for how the teams develop the concepts; however, 3P requires fast development of concepts and ideas in order to quickly kill them if they prove not to be viable. "Fail fast. Fail cheap," was the maxim espoused by Marie Elena-Stopher, our mentor and facilitator for many 3P events. Going fast requires very different approaches than an engineering team would typically force themselves to do. Often the assigned functional alternatives will be a mixture of known and understood concepts with some very new approaches. During the discussions that take place as the team copies the Process at a Glance steps onto their worksheets, they will flesh out what they know and what they need to investigate. It is worth assigning a risk factor to each of the given process steps, perhaps simply rating and numbering them sequentially based on lowest to highest technical risk. Other rating processes could work also, but this step in the 3P event begins to take the form of a mini-project management process. Limited time and resources make every decision count. Always tackle the highest-risk items first to avoid getting caught short of time later. Keeping the event targets visible helps the team keep those criteria in focus as they develop their designs.

Referring back to the *Shewhart* (or Deming) *cycle* of Plan, Do, Check, and Act, the team is now squarely in the planning stage. They must investigate and research the areas in question, begin preliminary layouts for their operation, and consider the elements of flow, transitions between steps, and the other criteria that need to be incorporated in their design. A scaled paper sketch of the intended prototype design should be done before breaking out the hammers and nails. This is not necessarily a 3P guideline, just basic carpentry.

Very quickly the team will move into the "Do" portion of the cycle, and it is typical that some portions of the team quickly start building while others are assigned to continue the research. As each aspect of the new operation prototype is set in place, there will be opportunity for the team to see and touch it, making evaluations and conducting an informal "Check." This brings them to the "Act" portion of the cycle, and the inevitable adjustments that get made to fine-tune the prototype concept. As the 3P event goes on, the teams will continue to refine concepts, sometimes abandoning initial approaches in order to move to better ones. Shewhart's cycle continues to spin as new ideas get created, implemented, reviewed, and improved at a velocity rarely achieved by other means.

Researching New Concepts

Developing new ideas and technologies can be very time consuming. The Internet provides a lot of the information needed to develop the concepts and will point the team toward subject matter experts who can help. Individuals on the team can take on specific modules or aspects of a module to research. They must determine technical feasibility and how to apply the methodology to the actual design.

Equipment vendors offer technical support and will often provide consultation to the team at no cost. Many are willing to come to the site and conduct demonstrations of the equipment, set up visits to see the equipment in other plants or their showroom, or provide on-line product demos tailored to your project. When presented with the urgency of need for this information, many vendors will rise to the challenge, to help you.

Raw material suppliers are also a good source of information. They handle their products all the time and may have knowledge of how other customers are using the raw material and moving it around in their operations. They can provide insight into handling, storage, materials of construction, environmental, or personnel safety issues and a variety of information specific to their materials, and often not made public due to its relative confidentiality. Gaining an understanding of material specifications can also be invaluable. Often we make assumptions on supplier capabilities and either overspecify, adding cost, or underspecify, permitting variation. If they have larger customers, sometimes a better grade costs less due to economies of scale. Interacting with vendors applies to raw materials, packaging, and other consumables that go into the manufacturing of your product. Do not overlook this source of information or assume that a change means added costs. A phone call or actual visit to a nearby key raw material supplier will almost always enable some better ways of working that can have benefits for both parties. Note that not every detail of the purchased materials needs to be pursued during the 3P event. Most can be done before or after the event. Pick the ones that you believe will have the greatest impact on the work developed during the event.

Internal subject matter experts should always be used when available. They may have direct knowledge of the new technology being researched or an understanding of how the new technology could affect the product being manufactured by anticipating side effects of the change. As an example, changing from convection heating to microwave heating will both increase the temperature of an affected object, but the physical mechanism employed to do so is radically different and could have unintended consequences. An equipment vendor would have a good understanding of how to apply the technology but might be unaware of possible product impact caused by the technology. An internal subject matter expert, familiar with the product, is more likely to anticipate those impacts.

Every industry and many companies will have literature and databases available. User groups and public forums may be available as well and might be useful if the information needed is not confidential. The team will likely have many ideas on where to go and who to ask in order to get the application information they need to develop their design ideas.

Identifying Risk

Every change involves potential risk. The risks need to be identified, assessed, and eventually either mitigated or accepted if they pose little impact to the project or product. The challenge in 3P is to identify the significant risks as quickly as possible, and to record medium to low risks so that they can be addressed at a later time. The large risks need to be flushed out and understood well enough to enable a decision. Is the potential value worth pursuing given the effort that might be needed to mitigate the risk? Can the risk be reduced? How much effort will it require?

There are many methods of evaluating risk including Failure Mode and Effect Analysis (FMEA), Hazard Analysis, New and Altered Analysis,

Management of Change, and many others. There are company regulations and governmental regulations requiring that the impact of change be understood and documented. This is especially true in the areas of environmental, safety, and quality. As the 3P design process develops beyond the event, for many industries, those considerations and analysis will need to be completed. This level of detail is prohibitive during the prototype phase of the 3P event. It is necessary to consider the risk imposed by the desired change, relative to the high-level impact.

Could the proposed change hurt anyone? Could it have any environmental impact? Could it cause regulatory problems? Could it have a negative effect on the product? How easy will it be to nullify or marginalize those negative impacts? Having the team make these assessments quickly and with open minds is often all that is required to continue developing the concept. If the response is negative or the challenge is simply too difficult, the team must make the decision to abandon the idea and move on, documenting the reason for the change in direction. If in doubt, these are often the types of questions that the Event Judges can help with.

Constructing the Three Prototypes

Teams always enjoy physically constructing their prototypes. They get to unleash their creativity and use hand tools and power tools to physically build mock-ups of their design ideas. They become competitive with the other teams, their energy level goes way up, and they really have fun doing something that has significant future benefit to the company and is very different from what happens in their "day jobs."

Concept Viability

At this point in the 3P process there is no need to build detailed emulations of the actual components. There is no time for that and the purpose now is to demonstrate viability of the concepts first, building in process flow, quality, and takt time for the operation being designed. There will be time to add detail and features at every progressive step of the Lean 3P design process. The challenge is always to enable the critical decision to abandon a concept or dramatically change it at the earliest possible moment.

Teams can get very creative when building their initial prototypes using whatever materials they have available to demonstrate the concepts. Assembly lines built on folding tables or 1 × 3 sticks work perfectly fine.

Figure 10.2   Common materials convey complex concepts to increase understanding.

They don't need to support any weight or be structurally sound. Duct taping a lightbulb to a piece of cardboard might simulate a sensor or the lamp in an infrared oven. Conveyor belts formed from a sheet of corrugated cardboard and marked up with a Sharpie to indicate the positions work just fine for visualizing the concept. Add some of the parts being conveyed by hot gluing them onto it, and a clear picture has been created. (See Figure 10.2.)

Often plants have equipment that is either not being used or is obsolete and awaiting disposition. These "bone yard" items can be very useful in developing the prototype concepts. The downside of using equipment that is still in service to Operations is that it will likely need to be redeployed into service when the 3P event is complete. If this equipment is used in the final combined prototype, it will prohibit the model from staying up intact after the event. This trade-off should be evaluated, though anything that can be done to help the 3P event participants understand the design better should be given priority. Developing a deep understanding of the concept enables effective decisions on viability.

Each prototype team could have from four to eight members. A natural competition occurs between teams, but at times a member on a different team will have a specific expertise in something another team is applying. When that situation arises it might make sense to have the expert collaborate with the other team to provide input. This will not likely detract from what their team is doing and can be of real benefit to the team needing the advice. Good-natured competition can be fun and add some motivation to the team effort. When borrowing talent from another team, try to minimize any negative impact to them by making good use of the time taken from them. Accept the advice on their timeframe to the extent possible. Always keep in mind that no "one" mock-up is selected for the final design. This makes it critical to vet ideas and test viability in order to thoroughly explore the alternatives. One hybrid model made up of the best features from all three models will be built in the end.

During the pre-event planning, common materials and tools will be listed. These items need to be available in good supply in order to keep the 3P prototyping phase moving ahead quickly. Nothing is more frustrating than watching a team wait for a saw or a glue gun while another team tries to finish up what they are working on. Have enough material and tools available to do the event effectively. Unused materials can always be returned to a building center afterward. The prototype build phase normally starts one day and finishes up the second. Do a quick evaluation at the finish of the first day and make arrangements to augment materials or tools as necessary before the teams arrive again in the morning. This is just one more example of the practical application of "Plan, Do, Check, Act," throughout the 3P event.

Safety and Power Tools

Mixed teams of people will have varying levels of skill using tools. This will range from some who are not sure which end of the hammer to use, to others who are skilled craftsmen. Tool safety is critical, and it becomes even more critical when using power tools. When building mock-ups and prototypes always use the lighter boards unless a heavy structure is required. Lighter materials generally cost less, require less cutting, and are easier to assemble. Do not use plywood when corrugated sheets or foam board will suffice. These choices help minimize the need for power tools.

Power tools should *only* be used by skilled personnel, and *only* with proper safety equipment. Some locations and some companies may require special permitting or pretraining before using power tools. If in doubt, seek council from your company's legal or insurance expert to understand and mitigate any liability issues ahead of time.

Process at a Glance

As each team builds the various sections of their prototype, they should continually update their *Process at a Glance* worksheet. Each value-adding function should have started with a description of the Material and Method.

The *material* description will include a few action words describing what the process step is set to achieve. As the design effort develops the Material section should also address material variation expectations, defining tolerances and specification ranges that will be accommodated. This aspect of design is too often underestimated, and highly complex machines tend to require very consistent materials to work optimally. Understanding normal variation within the materials at the earliest point helps to build accommodation for that variation into the design at the earliest moment and the lowest cost.

Method includes a description or sketch of the equipment depicting how it will operate to accomplish the process step. As development continues after the event, the sketches will evolve into full computer-aided design (CAD) models of the concepts, complete with part numbers, dimensions, stacking tolerances, and other design criteria. Initially a simple description or sketch will suffice to convey understanding.

How will you assure that the process step was completed correctly and is defect free? Evaluating a means to *gauge* the process either manually or automatically should be included here if it has been determined. The means of gauging a process step can be as simple as having an operator inspect the parts on some given frequency, or bringing a tool over to verify a key measurement. If warranted, a more expensive verification method such as an on-line sensor or vision system might be incorporated. At this stage in the design development it is not expected to have great detail built into gauging. Giving thought to gauging or quality verification early in the design allows consideration for space requirements and relative complexity.

Does the process step require a special *tool* to accomplish it? Many operations require specific tools to complete them. A picture or description of a tool required should be featured on the chart. The types of required tools can be extensive and tend to be more prevalent on less automated operations. The required tool might also be a factor when considering layout and location. If an operator needs to use a tool on a high frequency, it should be located at arms distance from where he or she will use it. If the tool is only utilized once or twice per hour, it might be several steps away, but a fixed location still needs to be determined. Be clear if the tool will be a custom design or a standard off-the-shelf model.

What kind of *fixture* or *jig* might be needed to hold the part in place while it is being processed? If it is a process involving fluids, what type of container is needed to contain it? In most cases high quality with excellent repeatability will require some means of keeping the parts in fixed positions, but once the functional step is performed the part must move to another location. Designing a mechanism to temporarily hold something in place and then quickly release it to be transported can pose difficult challenges that the team must consider. Again, the details of the design do not need to be developed fully at this stage; however, the consideration should be documented.

Will this prototype step require any special guarding or containment for leaking fluids? What other *machine* considerations need to be incorporated at this juncture to protect the product or people working around it? Are the heights right to enable good ergonomic access? Will there be a means of lifting a heavy piece of equipment out to be maintained? Are there access panels located in the right places to service the machine or make required adjustments? Is there anything special to consider about the lighting or ventilation for this application? This section of the Process at a Glance is intended to capture important considerations associated with the method being employed in the functional step.

Overview Comments on Developing the Three Prototypes

All of these criteria should be updated during the prototyping operation as they are being considered. The very act of reviewing them and attempting to address the questions provides a strong *check* in the process and enables the teams to *act*, making required adjustments to improve their design.

It is important to note that this portion of the event needs enough time to "breathe." The groups must have the time required to fully understand each alternative in order to have confidence in it as a viable option. If this step is rushed in order to get on to the next phase, the teams will generally gravitate to the tried-and-true solutions, reducing risk, but also reducing the potential for the breakthrough ideas. Think about it. If the goal of Lean 3P is to rapidly increase knowledge and understanding to allow participants to take greater risks for greater benefits, limiting this step will close down that process. Set realistic time frames for alternative development in order to gain the most benefit from the 3P process. Normally this stage of the process will take a day and a half or longer to complete. It should not be cut short without jeopardy of abandoning potentially great ideas. Often teams will work long hours the first day of developing the prototypes, researching the concepts and arranging to have outside resources brought in to demonstrate equipment and new technologies.

The three teams must continue working diligently to complete their prototypes within the allotted time frame. In addition to building the physical model, they should complete as much of their Process at a Glance as possible and prepare for how they will present their design to the other participants and Judges. There will be time limits for the report-out session, so developing a succinct presentation that provides enough detail to be understood is the goal. The other participants and Judges will be able to walk around the prototype model, touch it, and ask questions, so unlike traditional concept reviews, the 3P evaluation becomes a powerful means of conveying the total concept in a very tangible manner.

There is still a likelihood of product changes during the prototyping phase of the Lean 3P event. You will still be learning all kinds of things about the process and by default the product itself. As your understanding of the process capabilities grows, there will often be product design changes suggested that will increase the synergies between product and process. You cannot plan for it. Just be open to the possibilities, and continue exploring the benefits of concurrent design.

Example of Developing the Three Prototypes—Kathryn's Finest

The 3P participants from Kathryn's Finest had been divided up into three teams. The Facilitators had put up flip charts in three sections of the workspace with the team members names preselected on them. Assignment of the better three alternatives was done quickly and with almost no discussion. In several cases the functional alternatives to each team were given to a specific team because a member of that team had a strong understanding of it. The rest were distributed randomly.

They were given a safety talk and some instructions in how to use the various tools available to them and the required personal protective equipment needed. Other ground rules were also covered to set the stage for a productive and safe prototype-building phase of the 3P event. Figure 10.3 depicts the Kathryn's Finest team reviewing safety rules.

The first step each team tackled was to sketch out their initial Process at a Glance. They could fill in some of the Material, Method, and Machine sections with the initial information provided while developing the seven alternatives. The Operations Manager appointed a lead for each team and they each started talking through options with their teams.

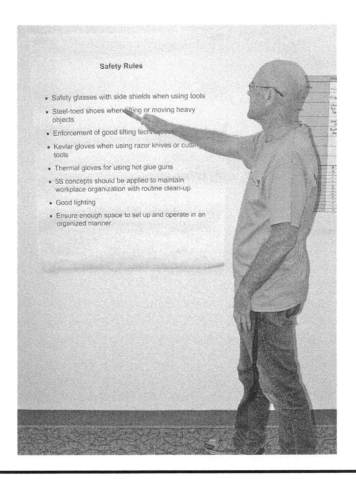

Figure 10.3   Safety is the first priority during a Lean 3P event.

Meeting overall takt time became a concern for all of the teams at some point in the discussions. The forecast for the new product looked pretty ambitious, and sinking a lot of capital into a major new operation seemed risky. One group was already planning to break the operation up into two or more smaller lines that would cost less and could grow as demand was demonstrated. This allowed each line to run slower and seemed to have some advantages. In their scenario, even the required capital for building the next line would be on a "pull" basis, driven by actual customer orders. Other groups thought that it might drive up labor costs and were planning a larger, more automated operation.

The teams at times seemed overwhelmed with the number of factors they needed to consider. Could they meet the required takt time? Would they be able to pull material from the prior stations without needing large buffer inventories? Whole apples were not consistent. They came in different shapes and moving them around without damaging them would be difficult. How would the line flow? How would the apples and other materials be advanced from one station to the next? Would they ever be able to put this all together into an actual "system" that brought all of the necessary criteria together in an effective manner?

As they started to understand the ramifications of actually using their given alternatives, reality struck. This was going to be very hard, and maybe impossible.

Over the long hours ahead they took each issue and conducted research, gathering information, sketching out rough designs, and physically creating mock-ups of their ideas. The team leads helped focus the activity and had the team create a list of key issues and risks. Those were quickly force-ranked, and a prototype project plan was developed.

For each team, the first few hours brought both excitement and some fear of failure. The Process Owners and team leads stepped in as needed to encourage the participants and demonstrate confidence in them and in the 3P process. They *would* be successful.

As the planning and information-gathering phase provided understanding to develop rough sketches of their prototypes, some team members started building them, while others continued to research. Everyone was fully engaged in the work.

■ Section Summary

☑ Form Three Prototype Teams
☑ Safety Talk
☑ Start the Process at a Glance

Baked Apple Project—Team 1's Preliminary Assessment

Team 1 gathered around their Process at a Glance diagram (Table 10.1). One by one they went through the value-adding functional steps and discussed the things they knew, along with questions.

The team lead, Philip, first wanted to develop the overall strategy for the operation. They looked at customer demand projections and the takt time calculations. Most of the team thought that they should try to build a single line that would handle a third of the demand, allowing them three times the time to perform each function. A couple of people were reluctant to go in this direction, thinking that it would drive up labor costs and capital costs with three lines to build, operate, and maintain. Eventually they agreed to go along with the majority, with some reservations.

The team agreed to build material flow into the process, recognizing that there would be many challenges. One of the team members wanted to keep it as simple as possible to keep costs down, but the process design engineer was hoping to fully automate it and keep labor costs down. He struggled to see how overall Takt time could be met without automation.

Table 10.1 Team 1's Preliminary Process-at-a-Glance

Team 1 Process at a Glance

Process step	1 Accumulate	2 Separate	3 Wash	4 Dry	5 Make hole	6 Fill	7 Heat	8 Cool	9 Protect	10 Identify	11 Group
Material	S.S. Hopper	Pick and place	Submerse water wash	Infrared heat	Drill	Meter in with auger	Deep fry	Flash cool	Clear clamshell	Stamp image	Stretch wrap
Method	Storage bin with star valve	Vacuum lift and servo drives	Barrel washer	Infrared heat lamp system	Auger-style drill on servo	Powder filling auger	Custom deep-frying equipment	Liquid N2 flash cooling	Clamshell packing system	Stamping machine on pre-printed shell	Place and stretch wrap equipment

The Accumulate function was to use a stainless steel hopper that would recycle back and forth between the supplier and them. It seemed expensive, but the recycling aspect of it had appeal, even though tracking the hoppers, keeping them clean, and getting them back as required would be difficult. If selected they would need to develop a system for managing that. They also struggled to see how a star-valve on the bottom would work to meter the apples out without bruising them. Philip made a note on their "bucket list" to call Jed at the Farmer Brown Apple Supply Company to get some technical help.

Applying a "pick and place" unit that used a vacuum system attached to a moving arm to separate the apples did not seem too difficult. The technology was common to them in previous designs.

Washing using a "barrel washer" was something new to them. They normally got fruit for their products already cleaned and prepared. Dealing with fresh apples and having to clean them before preparation was foreign to them. During the development of the seven alternatives, the group that had suggested it had shown a Web site with commercial barrel washers that were designed for apples or potatoes. Philip made a note to have someone call that vendor.

The fourth step, drying the apples using infrared heat lamps, seemed a bit daunting. The concept was simple enough, but how would they prevent the apples from being overheated before they were ready to be baked? They agreed to come back to that point later. The concept was reasonable.

Making a hole using an auger sounded easy to the group, until someone reminded them that the hole had to be in the right place to fully extract the core. Fixturing it properly would be a critical factor in this part of the design.

The cooking step was one of the more unusual alternatives. Using a deep-frying technique to cook the apples in melted butter seemed possible. One team member suggested putting the brown sugar right into the butter mix to flavor it and sweeten it at the same time. This was controversial, and Philip noted that a first step would be to get the food laboratory working on some prototypes of this proposed process. This also meant that the previous step, Filling, would not be needed, if the sugar and melted butter combination worked. One of the team members took exception to this. She felt that if the process required 11 steps, than they should not arbitrarily eliminate one. This caused quite a lot of debate and more than a few heated comments were made. Philip suggested they take a short break and talk about it afterwards. During the break he consulted with the facilitators, and with their endorsement, the team was able to reach consensus. They had the freedom to change the order and approach they took with their alternatives.

The rest of the planning session for the prototyping phase went fairly well. Flash cooling was already being used in another part of the plant. Packaging in a clamshell was not expected to be too difficult, and identification using a stamping mechanism on a preprinted label was also a common practice at the plant. There was some concern about stretch-wrapping loose packages, however, and a couple of the team members were pushing back on the approach, suggesting some new ways of achieving it that were more commercially viable.

Controlling the process was still a big question for the team. Determining how people would interface with it, or even how many would be required for each of the lines, still needed to be determined.

Team Lead Philip worked with the group to develop a plan of attack, and with clear expectations set, Team 1 was off and running.

The other two teams were also progressing in a similar path forward, with some very different alternative considerations. (See Table 10.2.)

- ■ Section Summary

  - ☑ Team Lead Develops a Plan
  - ☑ Healthy Conflict
  - ☑ Consider Takt Time, Pull, and Flow
  - ☑ Develop the Design

Process at a Glance, Selecting Teams, Constructing the Three Prototypes Recap

- ☑ Divide the three better alternatives among the three teams.
  - ☑ Select team leads for each team.
- ☑ Each team furthers the ideas using Process at a Glance.
- ☑ Begin prototype construction once ideas and connecting ideas are well thought out and understood.

Table 10.2 Starting-Point Process at a Glance Charts for Teams 2 and 3

Team 2 Process at a Glance

	1 Accumulate	2 Separate	3 Wash	4 Dry	5 Make hole	6 Fill	7 Heat	8 Cool	9 Protect	10 Identify	11 Group
Process Step	Accumulate	Separate	Wash	Dry	Make hole	Fill	Heat	Cool	Protect	Identify	Group
Material	Horizontal feeding	Robotic arm	Spray water wash	Forced air	Punch out	Squeeze in	Microwave	Cold air	Sealed box	Printed label	Bag
Method	Rotary pack-off table	Teachable robot with suction lift	Low-pressure spray wash	Air knife system	Punch press	Extruder	Microwave oven	Blower with cold air	Serving bowl in box assembly	Print and apply system	Auto-bagging system

Team 3 Process at a Glance

	1 Accumulate	2 Separate	3 Wash	4 Dry	5 Make hole	6 Fill	7 Heat	8 Cool	9 Protect	10 Identify	11 Group
Process Step	Accumulate	Separate	Wash	Dry	Make hole	Fill	Heat	Cool	Protect	Identify	Group
Material	Bulk bags	Gating system	Scrubbing system	Warm air	Knife cuts	Pick and place	Convection	Freeze	Microwave bag	Inkjet	Corrugated shipping carton
Method	Bag dumper and filler	Vertical slide gates	Rotary brush washer	Blower with dry and warm air	Rotary knife with scoop	Pick and place frozen blend	Convection oven	Place in freezer	Bagging/sealing equipment	Commerical inkjet printer	Auto-cartoner

Chapter 11

Rapid Evaluation of the Three Prototypes, Selecting the Final Prototype

Introduction

The primary goal of process design using Lean 3P is the overall system performance. Critical decisions, such as to build one process operation or five, have to take precedence over an individual alternative's beneficial attributes. There is a hierarchy of decisions that have to be considered. First we harmonize the product attributes with the overall process capability. Then we seek to optimize the overall process, looking at scale, levels of complexity, and the role and needs of the people who will interact with it. Finally we look to the sub-systems and evaluate the benefits of each facet, in harmony with the whole.

Rapid Evaluation of the Three Prototypes

The teams must have a hard stop for completing their models. Without a deadline, in theory they could just continue to progress their ideas through to final design. Having a hard stopping point forces them to prioritize and push to get far enough along that they can effectively present and defend their concepts. Every team will have some concepts that they favor over others. These people are subject matter experts who have designed or operated or maintained similar equipment before and have now been forced to explore new and different ways of accomplishing these value-adding functions. The fun comes when one or two of the concepts come through so strong and beneficial that the team has one of those "eureka!" moments and realizes a breakthrough in an idea that they traditionally would never have considered or supported.

With three teams working diligently on three completely different approaches to every process step, it is not hard to imagine the power of the 3P process. It is typical for each team to have one or two breakthrough ideas, and often more than that. There is excitement in the room as they busy themselves in final presentations of the new innovations they created and developed.

When the teams make their presentations and as the evaluations of each step in their process are made, keep the focus on performance against the agreed criteria. There will be enormous pride of ownership for the new ideas, and there needs to be a balance struck between celebrating the effort that has been made and the unvarnished facts regarding the strength of the design concept. This is not normally a problem, and a good facilitator will keep the good-natured teasing in check, while focusing on the strengths of the design element being considered. There have been cases where an individual gets so shattered because his or her concept was not favorably received that the individual became negative when evaluating the other two teams' prototypes. This is not typical but needs to be watched for. When the evaluations of all of a team's designs have been completed, with all of the questions answered, there should be applause celebrating what they have accomplished.

The actual evaluation process starts with a single team randomly selected to go first. Have the team provide an overview of their approach to the design. The alternatives each team developed were dictated, but there are often many other fundamental differences in how they developed their models. One team might have technologies that lend themselves to creating multiple smaller operations that collectively meet the required takt time, but enable effective scale up and scale down to accommodate demand cycles. Another team could have found a way for pulling in formerly disparate functions, separated by departmental silos or physical barriers to create process flow. Simply judging the effectiveness of the individual modules can miss the broader gains that were considered and developed.

Evaluation Participants

All of the 3P event team members and the Judges (Critical Evaluators) participate in the evaluation of the three prototypes. Although only one team presents at a time, all members get to discuss, ask questions, and ultimately determine the score of the individual process steps. There will be some bias toward the specific processes that the teams developed and that can provoke good debate and increase the understanding of the entire group. Ultimately the goal is to have all of the participants and Judges fully understand the benefits and trade-offs of the design so that a fact-based decision can be made for the chosen alternatives that will move into the final prototype and design.

Process at a Glance and Complete Process Overview

The team presenting their prototype should first reference their Process at a Glance and describe the alternatives they were assigned for each function. What were they? How did they approach it? What resources did they use to verify the feasibility of their design? Does every functional step in their process meet the required *takt time* inclusive of the transition steps? The presentation should quickly cover all of the process steps focusing on the method used and the approach taken to apply it. This will give the others an opportunity to better understand the strength of the concepts and gauge the amount of risk that might be involved.

The team should also document their overall approach to the operation they designed. This can be represented on the Process at a Glance sheet in a last column. How did they incorporate flow? Did they create one large operation or several small ones? How much automation was used versus operator involvement? During the interview they will explain to the others how and why they approached the design in the manner they did.

Congregate around the Prototype Being Reviewed

Armed with a general understanding of the concepts, the group will then move to the actual prototype model where the presenting team will take them through their design, module by module. They should point out design criteria that might be significant and noteworthy at each step in the process. Does it enhance flow? Did they incorporate a gauging mechanism or Poka Yoke? Did they use low-cost automation to make a process step simple and inexpensive?

The prototype design team should also focus on the transportation and transition elements of their design. These are generally not covered in the Process at a Glance at this stage but are important because of the potential costs and complexity they add to the design. Often a team will develop a very novel approach to moving or inspecting in-process parts or materials that will end up being incorporated into the final design.

As the presentation progresses through each process step, the power of the 3P prototyping phase becomes hugely apparent. No one is left out due to misunderstanding. People can see and touch the concept. They can walk around it and see where the tools will be stored and how far the operators will need to walk to gather additional materials. While evaluating the mock-up, the group will gain an understanding of material staging, and how man, material, and machine will work together in the system. Potential problems will be highlighted as the group physically inspects the model. They can start to appreciate how problematic that column will be in the middle of the floor in the selected area, and how the team addressed it. The size of the space can be visualized allowing the group to fully comprehend the possibilities and limitations. This impact could never happen under the best conditions using 2D or even 3D CAD modeling.

Process at a Glance and Evaluation Criteria

After the walk-through of the team's prototype, it is time to score each method employed using Process at a Glance and the Evaluation Criteria developed on Day 1 of the 3P event. Table 11.1 represents a possible template to use, with the evaluation criteria kept on a separate board if space is an issue. As each criterion is about to be referenced, go over the definition so that the entire group is grounded in what they mean. This takes a few extra minutes up front but saves a lot of time in discussions later, and the teams generally reach consensus much more quickly.

To keep the scoring simple, again use the Pugh method. Compare the design being evaluated to an existing, similar operation, if one exists, or the best *known* method. This is the same method recommended for determining the better three of the seven alternatives. Use a weighting factor to give some evaluation criteria more emphasis. As mentioned before, a weighting factor of 3× can be accomplished using from one to three "**+**" or one to three

Table 11.1   Process at a Glance

Process Step		*1*	*2*	*3*	*4*	*5*	*6*	*7*	*8*	*9*	*Complete Process*
Material											
Method											
Gauge											
Tool											
Fixture or jig											
Machine											
Evaluation Criteria	*Weigh Factor*										
1. Minimal capital											
2. Changeover time											
3. Low-motion waste											
4. Pull system											
5. Autonomation											
6. Off-the-shelf equipment											
7. High process capability											
8. Internal waste collection											
9. Simple as possible											
10. Maintenance free											
11. Scalability											
Total +											
Total −											
Rating total											

"–" and is easy to accommodate and score. Score the process step as better than (+), about equal to (0), or not as good as (–) a similar process. As noted before purists of the Pugh method may wish to use "S" instead of "0" to designate a step being the same or equal to the comparative step. This evaluation process allows the group to move through the discussions and scoring fairly quickly.

When there is no similar or known operation to compare to, a numerical score may be beneficial. Again, keep the scoring as simple as possible. Score an amazing design as a 2, an acceptable design as a 1, and a lesser design as a 0. The key to all of this is differentiation in order to evaluate which of the three prototypes has the overall best concept for each of the alternate process steps. Some teams like to use numeric ratings even when comparing the new design to an existing one versus using the (+), (0), and (–) ratings. Either way can work. Avoid placing too much focus on the quality of the score and *do* place the focus on understanding and rating the relative value of the concept.

There is no need to add up the scores from the individual process steps because they will not be used as an overall comparison of the prototype. The final design will be a hybrid model using the best elements from each of these three mock-ups. It is worth taking some time, however, to take some notes on the overall approach the team used. Capturing the thoughts of the group after reviewing the prototype can be very useful before moving on to the next prototype evaluation. Some modules might only work effectively with other modules, and splitting them up will create suboptimization.

Define the aspects of their whole system that were especially impressive. Always work from the overall system level down to the modules, capturing the people impact at every juncture. Note the methods of transportation or transition that really stuck out as brilliant concepts. Understand why they designed for several smaller operations versus one larger operation and quantify the expected benefits and risks. Their strategy could enable a quick ramp-up of product capacity with lower initial capital cost, less space, and greater labor efficiency. Taking some notes at this juncture will complete the evaluation and capture all key lessons learned for future application.

Before moving to the next team, take the time to applaud their efforts and acknowledge the hard work each person contributed. This will take only a few minutes, but it sets the tone for subsequent presentations and honors the great work the team accomplished.

The entire group now moves to the second prototype, and the entire process is repeated. Every 3P event will be different; however, this step of the

process could easily take several hours to carefully go through each prototype and reach a decision on what to include in the final model. Balancing
the time required to understand each design concept with the time pressures
to move into the next step is challenging. The report-out time for each team
should be set ahead of time so that they can adequately prepare their presentation, allowing significant time for questions and dialogue.

Take Some Pictures

During the 3P event it is helpful to take pictures of the progress being made
by the teams. These are great for team building later and for reconstructing
what happened during the fast-paced event. More significant though is to
take pictures of each of the three prototypes. These prototypes will be taken
apart after the evaluation is complete and incorporated into the final design.
There might be lessons from these designs that will be lost once disassembled. Taking pictures preserves the learning and captures what the team
developed.

Selecting the Best Process Steps from the Three Prototypes

System Impact

After the third team completes the evaluation, all of the participants must
compare the best attributes for incorporation into the final design. Start
with a review of the highest-level system strategies and work your way
down through the system detail. When the teams have proposed funda-
mentally different strategies, compare them, and understand the benefits
and the risks. Then determine the optimum approach based on the Event
Boundaries and Goals from your Event Charter. Consider the risk involved
with the proposed system-level strategy, and if formidable, evaluate how
readily the risk could be mitigated. Starting with the overall system strategy,
based on a single model, could limit the applicability of some attributes
of the other models. This does not mean that you would ever adopt one
overall prototype of the three holistically. It simply means that the overall
system design is viable, and the benefits should be attainable based on the
model that was developed.

People Impact

With a system-level decision made that meets the Event Charter conditions, the focus should focus on people interactions. Evaluate the three models and look specifically at the role of the people who interface with it. Operators and maintenance personnel should be given top consideration due to their intimate interaction with the operation. Determine if the operator interactions in the one model are more engaging than the others, and understand why. The technologies employed might be easier to operate and repair on one prototype, compared to the other two. Get feedback from the operators, mechanics, and material handlers who will interface with the new operation. See what they liked about the overall system-level interactions on each prototype. The balance between people involvement and the level of automation can dramatically change the complexity and cost of design. Automation will add development costs (resources or time) and will likely enable lower operating costs and higher consistency in product quality. More manual operations swing the balance in the other direction.

Value-Adding Function Selection

Now compare the scores for each process step from the three models. It helps if the evaluation sheets are in close proximity when doing this. Compare one value-adding process step at a time, comparing scores, attributes, and risks. The entire group must select the better one of the three options to be included into the final prototype. Note any synergies between process steps or between a process and transition step, so that you do not suboptimize. As expected, the evaluation phase can generate some heated debate when the different options are reviewed. Every design option will have trade-offs to consider. The risk–reward ratios will need to be balanced, but eventually only one approach must be selected. If at a later date, after additional development of the design, a selected idea fails, the alternative second choice can be pulled and utilized. When there is more significant risk, two alternatives can be pursued in parallel for a given value-adding function. Keep the focus on the primary, but provide enough resources to the alternate so that it can be quickly substituted if necessary, and still maintain the project launch schedule.

The teams who developed the concept will generally be partial to their own designs. The evaluation process utilized as each of the prototypes is reviewed will generate data and facts for open discussion and optimal selection to occur.

After each of the individual process steps is determined, the attributes of the *complete process* should be reviewed along with an overall system assessment. These concepts can provide some of the biggest gains and should be thoroughly evaluated before moving on. Sometimes a team will develop a very different concept that can revolutionize the entire operation.

Once the decisions have been made for the final prototype, the entire group will become one large team to assemble it. The physical assembly is generally made up of the various parts from the individual prototype models. Typically one of the three prototypes will incorporate a higher number of the final alternative selection and can be utilized to create the final, saving some time and effort.

Example of Evaluating the Three Prototypes— The Baked Apple Project

For the past day and a half the three teams in the Kathryn's Finest 3P event had been working diligently on developing their prototypes. Spirits were high and everyone had contributed to their final designs. The Judges had not been formally asked to join the event teams since the evaluation of the seven alternatives, but a couple had dropped by to see the progress. They continued to be impressed with the level of innovation and creativity being demonstrated by the teams. They were especially surprised with some unlikely candidates who rarely added anything significant to routine designs or problem-solving sessions but were now running with new ideas and concepts. Somehow 3P was unlocking a lot of hidden talents.

During the early moments of the prototype development phase, Philip, the lead for Team 1, recognized the value in having one of their Process Outsiders, Jed from the Farmer Brown Apple Supply Company, come in to discuss ways that they supplied apples to other customers. They supplied apples in bushel baskets, corrugated bins on wooden pallets, and recently had been filling bulk bags with apples. Bruising was a problem when handling apples, and there was equipment designed to minimize it. Kathryn's Finest wanted to get only the best-quality apples, and to limit the size of the apples to certain specifications. They agreed to a size that represented the median-size apple Farmer Brown grew in their orchards, and coincidentally it was just about the perfect serving size for a baked apple. There would be a higher cost for the apples due to the selection process required to meet this tighter specification, and Jed promised to get back to them with a new quotation. This registered a strong concern with the procurement person.

The cost of apples was a big part of their cost structure, and they needed to buy apples at commodity prices, not specialty.

While Jed was there, he also discussed the types of handling equipment typically used by suppliers to transport apples, separate them, and wash them with minimal bruising. He recommended several large suppliers who could give them more information or even arrange a demonstration. Although one team had initially asked the Farmer Brown company to visit, each of the teams had people join in on the conversation with Jed, so that they all could benefit and work from the same knowledge base.

That first day of prototyping was long and intense. The teams continued to develop their alternative designs, and there was a lot of excitement as they started to put the first parts of the prototype line together.

Research was being done to verify concepts and designs. The Food Test Lab was working diligently to test the different heating and cooling methods to evaluate the impact on product taste. They tested each of the three main cooking methods and did a design of experiments (DOE) exercise in an effort to optimize the variables and achieve the very best combination of taste, sweetness, mouthfeel, and aroma. The best candidates were created using traditional baking in an oven using a cupcake-style cradle, with a ratio of sugar to butter that favored the butter. Running a close second was the apple that was "deep fried" in the hot butter and sugar combination. It had excellent flavor and aroma, and the mouthfeel was significantly better than any of the others. Even its appearance seemed more appealing. The only downside seemed to be sweetness. They could not increase the sugar content in the hot butter mixture enough to make the apple really sweet.

Armed with samples, they came back to the 3P teams. Team 1, who were working on the "deep-frying" process concept, reacted with disappointment when they heard the taste test news. After the grumbling subsided, Carl, the Senior Design Engineer (and an avid cook), suggested they simply add the missing "fill" step in their prototype *after* the cooking step. The added butter and brown sugar would quickly melt in the microwave oven when the customer prepared it at home, creating a wonderful syrup to dip the spoonfuls of baked apple into. The team thought it was a great idea and so did the representatives from the Food Lab. They ran right back and started preparations to test the postfill concept, while Team 1 went back to developing their prototype.

Sales and marketing were looking in on the proposed packaging and identification options, attempting to gauge how customers would react to them. On each of the prototypes the primary focus of the 3P participants was in determining technological viability and the estimated cost of their designs, quickly followed by a focus on flow, layout, ergonomics, and the other evaluation criteria. Not surprisingly different people brought specific focus to the design. The engineers loved the technology. The operators were concerned about layout and the impact on how they would have to interface

Figure 11.1   Simple materials can be developed quickly to convey powerful concepts.

with the line. Mechanics were considering access points and reliability of the selected equipment. More than a few pieces of equipment were nixed after the mechanics shared their experience having to fix similar units. Everyone added something.

Sketches of the new designs were being completed for each of the prototypes using pencil and paper, with approximate measurements for noncritical dimensions, and more detail for the main modules. The groups were gathering on an ad hoc basis as needed to brainstorm ideas and get input. Immediately afterward, a quick sketch would be drawn up and handed over to the team to try-storm, by building it.

The use of common materials such as corrugated sheets, board stock, and other items found in building supply centers can convey very powerful concepts. An example of this can be seen in Figure 11.1.

As the three teams finalized their prototypes, they continued to add detail to their Process at a Glance charts and worked to ensure takt time was being met at every step in the operation. An example of a sketch of the bulk-bag loading system and the finished prototype can be seen in Figures 11.2 and 11.3. The Facilitators kept bringing them back to the Process at a Glance, stressing the importance of documenting their considerations for purposes of presenting to the others during the report-out sessions. The detailed Process at a Glance charts would also serve as a reference point for future efforts, and potentially as a fall-back option after the event should one of the selected designs not work out as expected. Figure 11.4 is a sketch of a barrel washer. In Figure 11.5 the team is shown building the prototype.

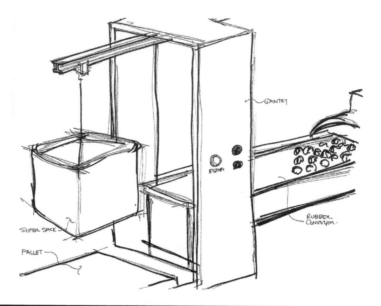

Figure 11.2 The bulk-bag apple loading system.

Figure 11.3 The bulk-bag apple loading system prototype.

Someone from maintenance suggested an access door to enable preventative maintenance activities. An access door is shown in Figure 11.6.

Outside of the event, the Facilities group had been working with another engineering company for design of the food-grade clean room facility. Several companies were being consulted, and they would provide

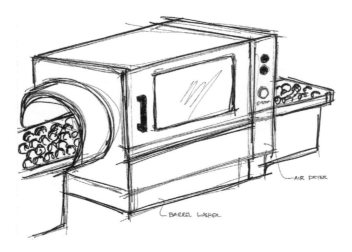

Figure 11.4   The apple barrel washer.

Figure 11.5   The apple barrel washer prototype in construction.

quotations before the end of the event week. This would be required for the final cost estimate and product cost calculations.

Finally the time had arrived for the report-out sessions. The Judges had returned and with the participants all gathered around, the Process Owners shared highlights of the prototyping experience. Much had been discovered and confirmed in the short prototyping period. The entire organization had learned more about the product and the operation they were designing than anything they had been exposed to previously.

Figure 11.6 The apple barrel washer prototype depicts a sliding panel for maintenance access.

- ■ Section Summary
  - ☑ Seek Supplier Input
  - ☑ Always Consider Product Impact
  - ☑ Sketch Process Concepts before Construction
  - ☑ Build Prototypes
  - ☑ Conduct Prototype Report-Out Session

TEAM 1 EVALUATION

The first team to present was Team 1. They were given a time limit and had been coached by the Facilitators on the presentation content and how to pace it, allowing enough time for questions and general discussion.

THE SYSTEM IMPACT

The group immediately observed that the size of Team 1's operation was small compared to the others. It utilized about half the space, and a general glance showed that the size of their prototype equipment was smaller than the others. Philip, the presenter, explained that because of the seasonal nature of sales, they had actually built a line that would only do about two thirds of the projected demand required, and that as the new product gained share in the market, they would build a second line to accommodate the growth and seasonal demand. He explained that this would also allow them to gain experience with the new technologies being employed and to modify them in the second line if required. Their projected capital cost for the line seemed about on track with the target, though everyone recognized the fact that only portions of the line would likely be chosen for the final design. Everyone agreed that it was an interesting approach.

THE PEOPLE IMPACT

Team 1 had a fairly automated process, explaining that because they were planning to operate two lines to reach the full demand projections, more automation was needed to keep the labor costs down. They were balancing the people aspects in other ways. Their prototype had a solid emphasis on the Maintenance Free criterion and had considered access points and materials of construction that would make it easier for them to make repairs and conduct preventative maintenance activities. The role of the operators was all intended to be value-adding, meaning that repetitive movements and more mundane tasks were largely avoided, but Philip admitted that they had not worked through all of the details on that, and their approach was not consistent.

PROCESS EVALUATION

Team 1 had worked with the Farmer Brown Apple Supply Company. Their design began with apples being delivered to them in medium-sized stainless steel storage hoppers that would be filled at the supplier and then recycled back and forth to them. The supplier could inspect and maintain them and liked the positive environmental benefits, compared to the disposable packaging they commonly used. Unfortunately the bins were very expensive and storage of both full and empty hoppers would require additional warehouse space, but because they were stackable, it seemed like an acceptable approach.

The apple hopper (function 1) would sit on top of a metering "star" valve, and a slide gate on the bottom would be opened releasing the apples into the first chamber of the valve. As the star valve rotated it would meter out one apple at a time onto a gravity roller where they accumulated at the end. At the end position a servo-driven pick-and-place unit (function 2) would use vacuum and a type of suction cup to pick up an apple and move it over in front of a camera. A motor would slowly rotate the apple, and the camera would take four pictures, enabling a vision system to inspect the apple for defects. Bad apples would then be moved over to a discharge shoot and into a pail below. Good apples would be placed over and released into a large tub-type piece of equipment.

The group evaluating the design was impressed with the environmentally friendly aspect of the hoppers, but there was no criterion selected for that aspect, so they made notes in the comments for this prototype. They scored the other aspects of both the accumulation system and the separation step before moving on to the wash system. Even though there were varied opinions regarding the design, the scoring for each criterion was straightforward and fast.

The participants evaluating the design raised immediate concern about using a star valve to meter the apples. One of the operators, who had experience with that type of equipment, called it the "apple-sauce" machine. That got a few chuckles, but the point was very real. The star valve represented a high risk for damaging the apples because of natural variation in the fruit. Philip took notes on the concerns, though his team had already raised similar questions. The participants also questioned the use of a vision system to inspect the apples. It would add a lot of cost and complexity, and the time required to "train" the camera software to reliably detect bruised apples was likely going to be lengthy, *and* expensive.

The barrel washer was a commercially available piece of equipment that Jed from the apple company had suggested to them for washing the apples (function 3). They had contacted the equipment vendor and learned that the standard units were much too big for their use, but the company was willing to make a custom unit on the smaller scale that they required. Their prototype built out of an actual plastic barrel was just about the size and shape of the one they would order if this alternative approach was selected. The participants really liked this concept. It was available commercially and seemed to have a proven track record.

Exiting the barrel washer, the apples would continue on a powered conveyor through a little tunnel that the presenter explained was an infrared heat lamp oven that they used to dry the apples (function 4). They insisted that they would use temperature controls to ensure that it would not precook the apples, but the participants evaluating the design had mixed feelings. Some felt that some precook would be beneficial and save time at the heating step. The operators asked what would happen if the line stopped and apples ended up sitting underneath the hot lamps. The saw the potential for many problems and did not think it was a great way to dry fruit. The design was evaluated using the rating criteria, and Team 1 moved on to the next function that had the apples accumulating and coming to a stop at the end of a gravity conveyor.

The next step in their process involved an operator who picked up the apples one at a time and placed them in a fixture that looked like little rubber fingers on a type of conveyor belt. They explained that coring the apple properly was a critical step and a major consideration for customer satisfaction as nobody wanted a baked apple with parts of the core left inside. A human could make sure that the apple was oriented properly to ensure that the core was properly removed.

When the apple had been placed into the fixture, the operator simply stepped on a pedal to index the conveyor to the next station where the apple stopped and a large auger came down to drill out the core (function 5). The team had developed an air blow-off to remove the pieces of core and apple and blow them into a bin adjacent for discard. They had even included a garden hose with a nozzle to depict a wash system for cleaning the discard bin after it was emptied.

After scoring the auger system, Team 1 presented their design for heating the apples. This step should have come after filling the hole with butter and brown sugar, but in Team 1's design they had to change the order of the functional steps. After drilling the hole, they had another operator release the apple from the rubber fingers and place it into a basket-style tray that had enough dividers to separate 20 apples. During the handling of the apples the operator could look at the hole and see if the core had been fully removed. Bad apples were tossed into a barrel next to the operator. When 20 apples were loaded into the tray, the operator pushed the tray down a short track where a hoist device lifted it and positioned it into a large deep fryer (function 7) that the team described as being filled with hot butter and melted brown sugar. The hoist was on a timer and would cook the apples for a set amount of time before extracting the tray and releasing it to the next station. Team 1 had actually gotten samples of apples made with this process from the Food Lab, and participants were given a sample to enjoy while scoring. All agreed it was slightly different than traditional baked apples but did have a wonderful taste. Figure 11.7 shows the sketch of the core removal process and the entrance to the hot butter station. Figure 11.8 shows the operator interface with the prototype.

The tray of apples moved on a conveyor into another chamber and the team showed how a door would come down to enclose them. Liquid nitrogen would be sprayed on the apples to flash cool them (function 8). The team showed a vent for exiting the nitrogen safely and other safety precautions needed to handle this material. This was another commercial unit that a vendor said they could make in the smaller size that Kathryn's Finest would want for this operation. The participants judging this design were impressed with the speed in which the apples were both cooked and then cooled, meeting the intended takt time with very small inventory buffers.

■ Although the apples might taste fine right out of the deep fryer, they already knew that it would not be sweet enough, so Team 1 included a step for filling the holes back up with butter and sugar. The tray exited the flash freezer and was mechanically pushed down another gravity conveyor to another station underneath a small hopper that was set up on a rail system. An auger inside of the hopper controlled how much butter and sugar mix was metered out into the apple hole (function 6), as determined by an operator who would position it over one apple at a time.

☑ Detailed Example of Prototype Report-Out
☑ Overall System Impact
☑ People Impact
☑ Process

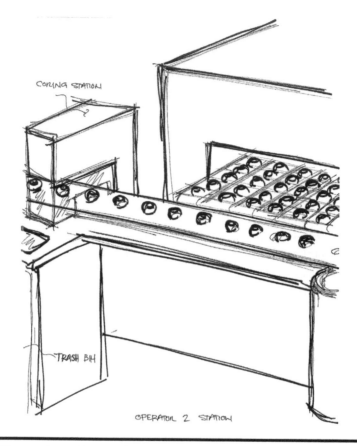

Figure 11.7   The apple coring station used to develop the prototype.

Figure 11.8   Operator interface with the prototype.

When all of the apples were filled, the operator pushed the hopper out of the way and removed the apples, placing them in an open clamshell that was on an adjacent conveyor (function 9). A dispensing machine took clear plastic clamshells and placed them one at a time on the conveyor. Each had a preprinted label affixed to the top and bottom. After loading the apples into the clamshells, the operator snapped the box shut and presented the bottom side to a stamping machine, which marked it with a lot number and a "use-by" date (function 10). All 20 boxed apples were then stacked into a corrugated tray on an automated stretch-wrap machine. The operator started the machine and the group of boxed apples was wrapped (function 11). Team 1 had also provided a vacuum assist system to lift the final package onto a pallet for transportation to the cold storage area.

The Judges and other participants completed the evaluation of the design and everyone applauded their efforts and innovation. They developed a very viable process. It had very strong people involvement, and the initial costs were estimated to be lower than expected.

Everyone applauded Team 1, celebrating the great work they accomplished over the past two days. A team photo had been taken earlier in front of their prototype, along with detailed pictures of their designs. Some of their alternatives would certainly be included in the final prototype design, but others would not. They recognized that having good documentation of their efforts might be helpful if the first choice alternative did not work out later, as it is fully developed.

Now it was on to Team 2.

■ Section Summary

- ☑ Review Prototype Details during the Report-Out Session
- ☑ Consider Overall System Impact
- ☑ Consider People Impact
- ☑ Evaluate Process Impact for Each Functional Step against the Agreed Criteria
- ☑ Have Frank Discussions to Learn and Understand
- ☑ Complete the Evaluations
- ☑ Celebrate the Team's Efforts
- ☑ Photograph the Prototype Designs

TEAMS 2 AND 3 EVALUATIONS

The evaluations of the two remaining prototype designs followed the same approach as Team 1. Each started with a high-level strategic overview of how they approached their prototype. Then they reviewed their approach to the people aspects incorporated into their operations.

Both of the remaining teams elected to make a single line that would satisfy the takt time demand. Team 2 applied more automation, and Team

3 elected to make their alternatives more manual and simpler. Team 3 also had the best flow, with a "U"-shaped operation that enabled operators to easily move to any constraint and help each other when required.

During the evaluations of each team's designs for their alternatives, debates were common. On a few occasions tempers flared when a concept scored lower than expected. This prompted more debate and discussion, and in every case a decision was reached.

A few concepts from the final two teams received very favorable response from the participants and Judges. Team 3 used a hoist system to lift bulk bags of apples off of a pallet and up above a conveyor, where a tie was loosened on the bottom of the bag to allow the apples to flow out. The conveyor then carried the apples past an operator inspection station where a gating system was designed to allow the operator to cull out any bad apples.

Team 2 had some great applications of Internal Waste Collection (one of the evaluation criteria that the other teams had struggled with). They also received high accolades for their drying system using an "air knife" process that looked very simple and effective.

When all three teams had finished their reporting they took a few minutes to reflect back on the highlights of each model. They had worked very hard and at a high level of creativity, with great ideas and suggestions coming from everyone in the group.

After a short break they would come back and choose the very best alternative concepts for inclusion in the final prototype design.

- ◾ Section Summary

    ☑ Complete the Remaining Prototype Evaluations

SELECTING THE BEST OF THE BETTER THREE ALTERNATIVES

With all three of the prototype reviews complete, the participants and Judges gathered together and discussed what they had just seen and heard. Considering the overall system first, they agreed with Team 1's idea of creating two lines that would share the overall takt time demand. Ideally, they wanted to design two lines that would each produce half of the demand. This created a lot of discussion because it was very different than how they had always approached new operations. However, they all learned a lot during the past 2 days and now felt optimistic that it was achievable, albeit with some risk. Fortunately one of the Judges really liked the idea and could see how significant the benefits were. Her support of the approach gave the other participants the confidence to voice their opinions and add to the support of the concept.

When discussing the people interactions with the new operation, there was no clear strategic direction from any of the three prototypes. Each had some exemplars of effectively considering and utilizing people. However, with the decision made to build two identical but smaller systems for the same total money or less, the people integration would need to be decided along with the selection of the best alternative designs.

Going through each functional step one at a time, they compared each team's approach, first looking at the overall pluses and minuses of each module and then developing the discussions in more detail. The Judges and other participants set up the three Process at a Glance sheets together on the same wall in order to do the comparisons more easily. They also took a number of trips over to the prototypes to see the physical conceptualization of the ideas. As they reviewed the alternatives, they recognized how critical it was to understand the concept as fully as possible before finalizing selection. There were elements of risk associated with process capability, concept viability, and capacity that had to be assessed at a deeper level, beyond pure reliance on the Pugh method scores.

For the first process step, everyone liked Team 1's "green" idea of using recyclable stainless steel hoppers, but the cost of building them seemed prohibitive. There would also be a disadvantage in developing a system for tracking the hoppers and ensuring that they were utilized effectively to avoid building up more than they needed. In addition their use of a star valve to meter out the apples seemed risky due to the potential of damaging the apples.

Team 2's use of a standard corrugated bin was very low risk. Using a standard lift table to dump them onto an accumulation table seemed practical, utilizing off-the-shelf equipment. There was some concern that this method might bruise the apples, and they acknowledged that it would need to be tested.

Team 3 used the bulk bags of apples, and they noted that these could also be recycled, but because it was a standard shipping package, Farmer Brown had already developed a system for returning them. The drawstring opening at the bottom of the bulk bag controlled the rate that the apples fell out and could provide more control of handling and less potential damage. It also allowed the apples to go directly onto a conveyor belt without a more expensive table design. The relative scores for the three alternatives reflected that Team 3's bulk bag and conveyor design was the most favorable, and it was the only design that had no minuses. This was the design selected.

The next function evaluated was Separate (the quality inspection to cull out any bad apples). Two teams incorporated vision systems that were commercially available; however, they were prone to technical problems and everyone felt that there was a probability of rejecting good apples. Team 3's approach of using a person to inspect each apple while loading it in the correct position for the washing step was very low cost. Although human inspection would not guarantee that every damaged apple was caught, the group felt that it was an excellent alternative as long as the operator could keep up with the process. There were also some ergo-nomic concerns that surfaced. However, the operators all agreed that it was not too different than many other tasks they routinely performed. They already had established ways of relieving each other to minimize repetitive stress concerns. The scores for Function 2 also favored Team 3's approach.

The rest of the evaluation went along the same way. Every team had some designs that seemed better than the others, and it was impressive to see how much research had been done in such a short time period. There were a few highly innovative design concepts developed, but the most inter-esting one was Team 1's use of deep-frying technology. It was not the least expensive process, but they demonstrated that it could handle the takt time requirements without the large inventory buffer required for the convection oven approach or even the microwave. They designed a long trough-like heater with the heated butter and brown sugar. Apples in stainless steel baskets would be lowered into the mixture and slowly pulled through using a chain-type mechanism. It was a custom design, but it looked simple, with low risk. Ultimately the decision was made due to the taste of the apple. This unconventional process produced a "baked" apple that simply tasted wonderful. When the added brown sugar and butter were added back and it was later microwaved, the added flavor created an amazing sensory experi-ence. Eureka!

The final discussion was around overall design criteria. Team 3 suc-cessfully selected many lower-cost alternatives for their designs, involving people at key functional steps and avoiding high-cost automation where possible. Each of the other teams had features that would be incorporated as well. The benefits of using people were significant, but the group ques-tioned whether people could handle more than 30 apples per minute and perform a quality inspection process. They would further study this poten-tial limitation as they developed the final prototype.

At the end of the session Nancy, the Purchasing Manager, came in look-ing furious. In her hand she held a quote for the "select grade" apples from Farmer Brown. The price had gone from a benchmark based on the com-modity price to a "commodity plus 30%" price. At that price, the project could never save enough elsewhere to make the required margins. Nancy was furious and the group was concerned that their hard work might have been just a waste of time. Alexandra, the Vice President of Operations, encouraged them to continue working while further negotiations were com-pleted and alternate vendors pursued.

The evaluations concluded at the end of the day, and most of the team members went back to take one more look at what they had created. There was a great sense of pride in those crude prototypes of cardboard, sticks, and carved up foam blocks. Each one told a different story and demonstrated a novel design that could have been progressed into a viable operation. Through the process they collaborated about pull and flow and maintenance requirements, throughput and costs, and how the design impacted each of the people touching it. They were tired, but the amount of progress that they made felt good to everyone. The concern about the apple price increase weighed on everyone, but Alexandra's assurance that it would be worked on provided some hope. Tomorrow they would all be working together on the final prototype and in developing the rest of the 3P project plan. Tonight they would sleep well.

- ■ Section Summary
    - ☑ Agree on the System Strategy
    - ☑ Agree on the People Strategy (if practical)
    - ☑ Determine the Best Alternatives

Table 11.2 shows the final alternatives selected by the 3P participants and Judges.

Summary of Progress

To capture how far the 3P event has come, consider that we started out with a basic knowledge of the essential value-adding process steps required to make the intended product. We then developed seven ways that those functions would happen in nature. Then each of those seven natural phenomena for each process step was converted into a technologically feasible approach.

The evaluation of the three prototypes typically takes place between the end of the third day and the middle of the fourth day. Going back to our graphic of 3P event phases, we are now entering the final phase— Optimization (Figure 11.9).

The *seven alternatives* were evaluated against the criteria, and the *better three* of the seven were selected to progress further. Those three alternatives were researched, and three prototypes were created.

In the last phase, the three prototype processes were thoroughly evaluated against the original criteria and event goals and from every process step the very *best* concept was selected for inclusion in the final prototype. Now the team is ready to build the final prototype and develop the project plan and information needed for the final report-out session.

Table 11.2 Final Process-at-a-Glance Charts for the Thee Prototypes

Team 1 Process at a Glance

	1	2	3	4	5	6	7	8	9	10	11
Process Step	Accumulate	Separate	Wash	Dry	Make hole	Fill	Heat	Cool	Protect	Identify	Group
Material	S.S. Hopper	Pick and place	Submerge water wash	Infrared heat	Drill	Meter in with auger	Deep fry	Flash cool	Clear clamshell	Stamp image	Stretch wrap
Method	Storage bin with star valve	Vacuum lift and servo drives	Barrel washer	Infrared heat lamp system	Auger style drill on servo	Powder filling auger	Custom deep frying equipment	Liquid N2 flash cooling	Clamshell packing system	Stamping machine on preprinted shell	Place and stretch wrap equipment

Team 2 Process at a Glance

	1	2	3	4	5	6	7	8	9	10	11
Process Step	Accumulate	Separate	Wash	Dry	Make hole	Fill	Heat	Cool	Protect	Identify	Group
Material	Horizontal feeding	Robotic arm	Spray water wash	Forced air	Punch out	Squeeze in	Microwave	Cold air	Sealed box	Printed label	Bag
Method	Rotary pack-off table	Teachable robot with suction lift	Low pressure spray wash	Air knife system	Punch press	Extruder	Microwave oven	Blower with cold air	Serving bowl in box assembly	Print and apply system	Auto bagging system

Team 3 Process at a Glance

	1	2	3	4	5	6	7	8	9	10	11
Process Step	Accumulate	Separate	Wash	Dry	Make hole	Fill	Heat	Cool	Protect	Identify	Group
Material	Bulk bags	Gating system	Scrubbing system	Warm air	Knife cuts	Pick and place	Convection	Freeze	Microwave bag	Inkjet	Corrugated shipping carton
Method	Bag dumper and filler	Vertical slide gates	Rotary brush washer	Blower with dry and warm air	Rotary knife with scoop	Pick and place frozen blend	Convection oven	Place in freezer	Bagging/sealing equipment	Commercial inkjet printer	Auto-cartoner

Lean 3P Event Focus Areas

Information	Innovation	Prototyping & Redesign	Optimization
Knowledge Gathering	Ideas and Alternatives Generation	Try-storming Convergence	Refinement Evaluation Planning

Figure 11.9   The final phase of the Lean 3P event–optimization.

Rapid Evaluation of the Three Prototypes, Selecting the Final Prototype Recap

☑   Evaluate each mock-up, process step by process step, reviewing system impact, people impact, and process alternatives.

☑   Score the prototype models.

☑   Review and hone in on one best hybrid final design.

Chapter 12

Developing the Final Prototype, Incorporating Evaluation Criteria, and Measuring Effectiveness: Final Report-Out

Developing the Final Prototype

A number of things must be done in constructing the *final prototype*. The first natural inclination of the group is to have everyone start tearing apart the individual prototypes and begin moving the pieces into place for the final. This needs to happen, but there is much more to be done at this phase of the 3P event. In the final leg of the event the prototype needs to address all of the key elements listed in the criteria and in the event goals. In addition a *cost estimate* must be developed, along with the *preliminary project plan* inclusive of resource allocations, key milestones, and timelines. Defining how the project will be progressed with new teams pursuing a variety of new directions must be developed to the point where requirements can be identified and committed to at the end of the event.

There is no formula for how to complete this last portion of the 3P event. The effort required will require skills from each of the functional groups represented. There are no longer three competing teams, but one large team, so organization is paramount. The effort must be brokered out based on key

211

knowledge and expertise. The entire team of participants will be available to work on this phase, so deploying them effectively will be a key to success.

Operations people and material handling specialists are generally good at ensuring flow, setting up supermarkets, developing changeover strategies, and identifying locations of tools. Engineers will need to develop the detailed cost estimates and resource requirements, assuring that takt time will be met in each functional and transition step, while achieving the required project delivery date. Supply chain or procurement people can begin developing their approach to outside vendors, suppliers, machine shops, and mechanical/electrical contractors. They can begin verifying lead times, determining how they will acquire bids and proposals, and establishing terms and conditions with new vendors. This is the point in the 3P event where all of the functional specialties, armed with a thorough knowledge of the project, must apply their skills and experience to create a final prototype and a plan for building the actual operation.

The functional steps selected for the final design will come from all three of the initial prototypes, and the teams that developed them generally understand them best. It is necessary to have those people involved in developing the cost estimate, required space needs, and other details about how their module will be incorporated into the new prototype design. The role of the event leader and the Facilitators will be to keep everyone engaged in the process and ensure that the most important details are developed by the right people.

Incorporating the Evaluation Criteria

The inclusion of all of the evaluation criteria is no longer optional in the final prototype. In the initial three prototypes there was such a strong focus on finding viable commercially available or developable applications of concepts that the evaluation criteria sometimes had to take a back seat due to the time constraints. In the final prototype all of them must be addressed. The criteria become a sort of checklist that every phase of the process must be verified against. Diligently going through each item ensures a robust process and forces the evaluation of trade-offs. How can we ensure that every function is executed perfectly in each step of the process and still meet a low capital cost criteria? What other options are there to ensure quality without use of expensive vision systems and sensor technologies? How could necessary personnel breaks be handled in load-load processes without adding operating staff? Will the mechanics require special skills to maintain this new equipment, and will they have ready access to aspects of the machine that will need routine service? How big will the supermarkets need to be to ensure supply to the line without requiring too much space or excessive travel by the operating personnel?

The number of questions and trade-offs discovered and resolved during this step in the process will be incredible. In traditional design, many of these questions will not be addressed until much later in the process, possibly via a shop floor Kaizen event well after launch, if at all. With the 3P process, most will be acknowledged and understood within the context of the whole project, allowing the teams to focus on the most important issues. It is helpful to have the teams keep track of these decisions and unresolved issues on a flip chart along with the names of the people involved. These notes can be used for discussion during the final report-out and may be beneficial for future reference. In the flurry of activity during this stage of the process, it is very easy to forget why some decisions were approached the way they were. Pictures and the notes from the flip charts can be *invaluable* later on in reconstructing the thought process.

Measuring 3P Effectiveness

Participants must complete as much of the work as possible ahead of the final report-out of the 3P event. Inevitably some things will be incomplete due to time pressures and the sheer magnitude of deliverables being worked on. The report-out information in order of relative importance includes

- ☑ Full construction of final prototype
- ☑ Layout with material and tool storage identified
- ☑ Process at a Glance (consider including takt time and cost for functions and for transition steps)
- ☑ Calculation of takt time
- ☑ Evaluation criteria assessment
- ☑ Event goals assessment
- ☑ Final cost estimate (include spend map)
- ☑ Build schedule and ramp-up
- ☑ Resource plan and teams

☑ Project timeline with milestones
☑ Spaghetti diagram of new operation
☑ Changeover plan
☑ Bucket-list and action items review

Each 3P event will have different needs, and final requirements can change. These deliverables are typical for many events.

Full Construction of Final Prototype

The final prototype is the main deliverable of the 3P event and the root that everything else springs from. This is done in conjunction with the new product, and even as the new production operation prototype enters the final optimization phase, the product is reviewed and can be improved. The purpose of the final prototype is to convey enough information about the design and layout to enable good assessments that allow the next steps of the design to progress with a high probability of success. As mentioned earlier, in large-scale processes this final prototype might take the form of a small-scale model of the process. This should still provide enough information for the team to approve the concept and allow the design to go to the next stage. When a small-scale model must be used, there will probably be an action plan that calls for the next generation of scale-up to be built in the days or weeks ahead, perhaps just focusing on the key modules.

The final prototype model allows people to see and touch and imagine how they will interact with the new operation. It includes how they will operate, resupply materials, utilize tools, make repairs, and complete paperwork or computer entries in the new operation. They will see where defective parts get rejected and be able to step through the changeover process. The final prototype is a valuable reference point that should stay up as long as possible after the event for the team to use during future stages of design. It is also a powerful instructional tool for others unfamiliar with the 3P process.

During the report-out session one or two people will step through the operation of the new process. This is especially effective when led by an operator, supported by engineering or other technical resources as necessary. It builds ownership in the new operation and recognizes the significance of each team member's contribution irrespective of position or title. Sometimes sharing responsibility for this walk-through step can be beneficial, allowing several people to participate and gaining the perspectives of an operator, a material handler, an engineer, or a maintenance person.

Layout with Material and Tool Storage Identified

The general layout of the proposed operating area has been a feature in each of the prototypes built and carries through to the final one, with the addition of even more detail. The prototype of the actual operation depicts how the team applied flow, gauging, and other major design elements. 3P considers the entire workspace within the scope of the project and attempts to define the locations of many other key items, ensuring total flow within the process area and maximizing the effectiveness of the people interacting with it, helping them do the right things routinely as part of standard work.

The floor should be marked off and where possible, physical arrangements made to depict where and how materials and tools are going to be stored in the new process.

If the operation will require routine changeovers to go from one lot to the next, it has to be considered in the development of the layout and the physical model. Changeover considerations include quick disconnects, pinning removable parts, identifying redundant parts that can be swapped out quickly, and many other equipment design features. It can also include locating changeover tools at the point of use, or developing changeover "carts" with all the necessary replacement parts neatly organized and with a place identified to store them that is readily accessible.

The final prototype should be laden with details that enhance performance and the people interface. The physical model allows us to physically step through key operating procedures, the changeover process, and material storage, access, and replenishment routings. Even with scale models, there is enormous value in walking through the people interfaces. Much will be learned in the process, and it will enable significant detail to be built into the prototype and operational layout.

Adding detail for these peripheral items makes it possible to visualize and physically step-off how far they are stored from the place where they will be utilized. This is a basic aspect of 5S (a five-stage workplace methodology involving sorting, setting in order, shining, standardizing, and sustaining), but unless you utilize 3P design, you generally only see consideration for the largest or most critical materials or tools, at best. That information would not normally be very important to a design engineer under deadline to get an operation built. 3P teams like to get specific with this type of information because it directly affects their jobs. They often include locations of major and supplemental raw materials, finished goods, required tools, safety showers, work supplies such as gloves and wipes, required personal protective equipment, and even locations for one-point lessons and access panels for utilities. Operators, material handlers, and maintenance people add enormous value here, utilizing their own experience dealing with previous oversights to ensure that they do not happen again. It takes 5S to an entirely new level with virtually no constraints to limit the group.

Process at a Glance

The Process at a Glance continues to be updated throughout the event and beyond. It is one 3P event element that needs to be continued throughout the life of the project. Not every aspect of the Process at a Glance will have to be completed by the end of the 3P event; however, enough must be done to affirm the directions being taken.

Every aspect of the Process at a Glance is revisited over and over. It embodies Plan, Do, Check, Act in a way that speeds through the steps until a satisfactory concept is identified that is worth developing to the next level of design. The teams develop their plans, build the model out of cheap materials, conduct mock-up tests with it, and then as necessary make changes to the original concept, improving it and pushing it through the next continuous improvement cycle.

Space limitations may not allow the physical prototype model to stay up after the event; however, the Process at a Glance captures most of the critical details. When coupled with good notes on the transition or transport plans and pictures of the final model, it becomes a powerful and visual reference point for the team going forward.

There is one significant variation to the Process at a Glance that should be considered. Most 3P practitioners tend to reserve Process at a Glance for only the key value-adding functions. At this phase of the event, and continuing for the life of the project, consider including the transition or transport steps. As noted before, the transition and transport steps in an operation add no value and can be highly complex and expensive, especially when they include gauging verification.

After the main process steps have been well understood and defined, consider adding the transition steps. Often this is where quality inspections are performed, verifying a previous value-added step was done correctly. Reject processes are built to segregate any defective product (only after mistake-proofing efforts are exhausted). Sensors are installed to verify and count defects, and software is written to measure and track them. Identifying the transitions and the additional functions that are taking place within them can be very helpful in managing performance and costs. When the design team has elected to use complex transitions, it is especially important to make visible the nature of the transition and inspection equipment. Including them prompts you to develop a better understanding of how they will function. What will they cost? How much material variation will it handle? What is their impact on takt time or costs?

This aspect of the Process at a Glance should only be considered after the main value-adding functions have been developed and optimized. In every case transition and transport of materials need to be quantified and well understood. They often represent an understated and significant risk to the project due to the lack of focus on it. The main functional steps will almost always get a strong review to verify capability. Transitions and gauging systems and the other subsystems that get placed between the functional steps might have a great impact on quality, costs, timing, maintainability, and other design criteria. Making transitions visible helps manage their risk.

A further addition to the Process at a Glance is to include takt time at each step of the process and for each of the transition steps. Recognizing the complexity of each step, including the transitions, is critical for success. Developing the takt time estimates for each of these forces the thinking process and highlights any areas of concern. Lastly, indicating the cost of each functional step and transition step can be insightful when added to the Process at a Glance. A rolled up estimate might not make visual the real costs of the pieces, and it is in these details that we begin to fully understand the project's complexity and risk. Understanding the costs can sometimes point a team to a simpler and less costly solution if identified early enough. There are significant benefits to adding the three elements of transition, function-specific takt time, and cost on the Process at a Glance.

Calculation of Takt Time

Takt time must be re-estimated for the final model to ensure that you have not overbuilt or underbuilt the operation. This should have been adequately considered at every value-added functional step already. Verify that the transition, transportation, and inspection steps can also achieve the required time limits so that the whole operation meets the design criteria. Do the staffing plans accommodate operating the line during breaks or meetings, or will that down time be built into takt as a hard limitation?

Always be brutally honest when estimating takt time on a new operation. Attempt to quantify the amount of risk and complexity being added. Equipment reliability can be greatly impacted from microstops. The relative complexity of the operation may create a lower utility due to a higher incidence of downtime while adjusting or replacing sensors, realigning parts, or replacing parts that wear too quickly. Maintenance teams or operators must have the skills required to keep the operation running at the expected performance rate, or there has to be a significant investment in training and perhaps an emphasis on higher-level standard work.

There are many examples where the side effects of a new technology can erode performance and jeopardize takt times. These risks and limitations can usually be overcome at some cost to time and resources. Always note your assumptions and be certain to identify and make visual those risks on the Process at a Glance.

Assessment of Evaluation Criteria

With the final prototype complete, the full assessment against each of the evaluation criteria must be developed. Some of the work done in evaluating the three original prototypes can be reused for the final. For the final design typically much more is now understood regarding the details of the chosen design, and care needs to be taken to fully evaluate it. Encourage the teams to ask the questions again, as things will likely change with the new configuration and additional information.

Scoring the final design against each of the evaluation criteria can be done using the Pugh method as described earlier. Does the new design seem better than the conventional approach, about the same, or not as good?

The checklist of criteria provides another great opportunity to assess the new process holistically. Each value-adding step of the process will be evaluated against the criteria, and then the total line should be judged as well, providing a general assessment and affirmation that the sum of the parts really did create a much better whole.

Assessment of Event Goals

Assessing how the new operation met the event goals is the highlight of the report-out session. It is the one aspect of the event that will get wide organizational recognition and acknowledgment. This is the summary of what was accomplished. It assesses the team's solutions to meet the required timelines and costs. It validates that the new product was developed to the extent required. It determines if the expected quality levels and throughput rates will be achieved. The takt time review determines if takt was achieved throughout the new operation or if elements of design were suboptimized, adding cost or affecting performance. The ongoing operational costs must meet or exceed expectations. *All* of the boundary conditions must be satisfied.

The event goals are always compared to the estimated results of the new operation. In addition to the comparison of the defined goals, you should indicate the other benefits developed using the Lean 3P process on the new operation. The assessment needs to show where additional flexibility was built in. It might show prototypes of new product features or enhancements. Equipment reliability issues should have been addressed proactively, along with other benefits, and they need to be captured.

The plan for achieving the target cost and timing is common to all projects. Communicating the expected ongoing use of stakeholders in the ongoing process of developing the product and new operation is not as common and may need more discussion. This extension of the 3P process, past the event, and throughout the entire project will require a commitment from the organization's management. After gaining experience with 3P, this notion should gain high acceptance based on the value it delivers. Initially there will be resistance to making that commitment. Companies that are used to running Kaizen events on a high frequency tend to think in terms of committing people wholly for the 3 to 5 days required for the event, but once complete they normally expect people to return to their jobs, with all of the major changes complete and documented. 3P differs from a typical Kaizen because it signifies the launch of a new product and process, not simply a change to an existing process as with most Kaizen events. Ideally this discussion will take place ahead of the 3P event; however, the plans and expected commitment levels required to go forward need to be communicated and accepted now that the scope of that need is understood.

The less-understood notion of launching a new product in a new operation and very quickly reaching a high standard performance level should also be articulated. How much time is anticipated to eliminate the constraints in the operation, create standard work for everyone interfacing with it, and reach the targeted standard throughput rates, while meeting the standard cost and quality rate targets? This often-overlooked goal can be the "elephant in the room." Everyone knows there will be start-up problems to work through, but what can be done to quantify the probable issues and mitigate them? It is strongly recommended to consider this and determine a target, understanding the effort required and the ramifications associated with attaining it.

Final Cost Estimate

Developing a cost estimate for the new product and project must be done for the final report-out. The product attributes and features have been defined, revealing great detail about material costs. The process is now being finalized as well, and the costs to manufacture, assemble, package, and test the product should be available to estimate. At this stage of development, many aspects of the design will still need to be defined and tested using actual equipment. Final costs will not be known for these aspects of the operation, and the accuracy of the estimates should be stated. Final specifications may still be in negotiation or evaluation. Other aspects should be much more accurate however, and for the final report-out it is better to have an estimate with all of the contingencies built in. The number reported will tend to be the one that is remembered, and no one appreciates when a cost goes up, even if the potential of it happening was explained. Cost is always part of the 3P event goals and will be compared to the original target. In almost all cases it should be less than the target or similar with a significantly greater performance level.

In general product and process estimates developed based on Lean 3P will be better thought out than with traditional approaches. There should be fewer surprises and less risk due to unknowns. When defining costs, state all assumptions. Will you utilize internal designers or an outside firm? How have you built in costs for internal engineering that should be capitalized?

Are you forced to be single-sourced on some equipment due to the technology selected versus bidding the work to several qualified competitors? What utilities will you need and what facilities' work must be done in the selected space? Be very clear what assumptions you are making to build your estimate. Changes will occur as additional design work is completed, and you must understand the risks to your assumptions.

If the estimates suggest that the product or process costs are coming in above target, there must be a concerted effort to examine trade-offs in either the product or the process, or some combination of both. Going back to the "must, should, could" assessment of product features and attributes, when faced with above-target costs it is fairly straightforward to begin eliminating "could" features and possibly "should" features to get into the target range. A similar exercise should be done with the manufacturing process steps. This type of challenge is never easy, but armed with great knowledge and understanding, a cross-functional team can make informed decisions for reaching the target goals.

If possible an estimated Spend Map should be included along with the final Cost Estimate. Phasing of cash flow is important to the business for planning purposes and should reasonably be estimated once the project plan and cost estimate have been developed.

Resource Plan and Teams

Closely associated to the cost estimate is the resource plan. A common pitfall for groups first using the Lean 3P process is to conduct an event and then develop the new operation the way you used to. The big ideas have been identified, and this fact alone will provide many of the ultimate gains. However, the cross-functional teamwork that created those big ideas must stay involved as the ideas are refined to ensure success of the complete project.

This demand on resources outside of Engineering is uncommon in traditional projects. The Operations team says what they want. Engineering goes off and builds what they think is needed, and Operations gets what they get. 3P is a very different model. Operations, Technical, Supply Chain, Finance, and all of the other groups are collectively responsible for building a new product and the associated production operation. To accomplish this successfully, they must participate frequently as the concepts are developed. Engineering might have the yeoman's task for doing the deep design, but the other stakeholders need to review and validate the changes and required trade-offs that inevitably happen. This is especially critical as design progresses. We continue to learn more, changes occur, alternatives do not always work as expected, final specifications change our criteria, and other issues develop. The ongoing cross-functional team is uniquely qualified to access the impact, develop new alternatives and options, and continue to maintain a strong alignment within the organization as the product and process ebb and flow on the path leading to the launch date.

This type of support from the other stakeholders requires a time commitment and a plan for how the interface will occur. Do not assume it will all work out later. Take the time to estimate the requirements for that interface to be successful and monitor how effectively it is working once it has been set into motion. Some organizations maintain a 3P team for the duration of a project, exclusive of other activities. Most organizations seem to approach it as a part-time activity, on top of their day job. This is the more difficult approach, but it might be the correct one for your project. Be as clear as possible when defining how much time is needed, and try to avoid the more ambiguous "20% of their time" estimates. Be as specific as possible as to when they need to participate in the process and what their role and responsibilities will be.

Some of this detail can be accomplished after the close of the event; however, the more that can be defined and agreed to during the closing meeting the better. People sometimes lose their sense of commitment to the new project when they return to their normal routines the next day. Getting a commitment in front of their peers and with their manager's concurrence can go a long way toward keeping them engaged in the months ahead.

Project Timeline with Milestones

Every project has a firm or implied deadline for start-up. This must be addressed as a fixed requirement and not allowed to go undefined until later. A danger with the new concepts and innovative outputs of a 3P event is the necessary time required to fully understand and develop new concepts and technologies. Inevitably there will be unknowns associated with these, and they will need additional study, testing, and development. Reliance on experienced vendors and subject-matter experts will help, but do not underestimate how long it will take to prove out a new technology.

Project management is essentially about risk management, and there are many excellent resources on this topic. For our purposes it is critical to identify where the highest risk is likely to come from and to define how it will be mitigated and when it must be complete. These milestones are essential for success. By mitigating the highest risk early in the project the probability of success goes up appreciably. Occasionally a new method will fail or require so much effort that a milestone cannot be reached. It might signal the need for additional resources to accomplish it in the required timeframe. Keeping the milestone visible might also suggest engaging a counterapproach concurrent with the desired method. Continue developing the desired method, but at the same time progress a tried-and-true method that can be substituted in time to meet the milestone commitment if required. This deviation from the original approved plan might not have the benefits of the intended methodology. When these changes occur, they should be done with an understanding by the entire team along with a revision of the cost and benefit goals.

In his book *Mastering Lean Product Development: A Practical, Event-Driven Process for Maximizing Speed, Profits and Quality*, Ron Mascitelli (2011) does a masterful job presenting techniques for quantifying the risk associated with new concepts, setting milestones, and managing complex projects visually and very effectively. One of the key concepts he develops is the idea of identifying the highest risks to the project and tackling those first with the idea that if most of the risk is mitigated early, the rest of the project becomes a tactical exercise with a high probability of success.

Spaghetti Diagram of New Operation

On the first day of the event, a spaghetti diagram tracing the movement of materials and people through the process may have been completed to graphically display the complexity and the amount of transportation waste. The 3P process should be a significant improvement over the traditional process, and developing a spaghetti diagram for the new operation and measuring transportation distances should show a dramatic difference between the two models.

Even if there was no existing process to compare to, and an initial spaghetti diagram was not done, it may be valuable to create one for the final prototype. Spaghetti diagrams demonstrate the flow of product, materials, and people, along with travel distances. They often enable further optimization of a process by making the path of operators and materials visible.

Changeover Plan

Many operations will have more than one product or variation of the product being produced on the same line. When switching from one product to the next a changeover process is needed. This can often be simulated using the final prototype model with operating personnel moving from station to station describing what they will do and how they will do it. Time requirements can be estimated for this activity. If the team desires they can do a dramatically scaled-down quick changeover analysis highlighting the key steps and attributes that were built into the new operational design. There are very good resources for learning how to do this Lean improvement process, starting with Shigeo Shingō's 1985 landmark book, *A Revolution in Manufacturing: The SMED System*, that defines the Toyota single minute exchange of die (SMED) process that has evolved into "quick changeover." The key elements for the report-out include identifying where the tools or changeover parts will be located to provide ready availability, what steps can be done ahead of time or afterward, and any factors built into the new operation that make changeovers fast.

A changeover plan is something that can be developed more fully after the event. Much of the benefit from quick changeover comes from first analyzing the necessary steps, organizing them to minimize downtime, and then developing standard work for all to follow every time it is performed.

The critical information required for the final report-out will be the time impact of the changeover and any assumptions regarding campaign length due to the impact on inventory levels. It will also be important to note the anticipated labor requirements associated with changeover. Often quick changeover analysis requires additional people to participate *only* during the changeover. Document all assumptions related to how this will happen, along with an action for completing the changeover plan later in the project. The beauty and power of Lean 3P is the ability to visualize upstream of all the downstream effects that the new design incorporates. The goal is to eliminate the need for conducting traditional quick changeover Kaizens after the launch, and to build in the features that will minimize or eliminate changeover impact.

Review of Bucket-List Items and Action Items

The summary of next steps in the report-out should include some discussion of the Bucket List items captured during the event. Bucket List or Parking Lot ideas that come up in almost every Kaizen event are ideas, concepts, or actions that may have merit but do not fit well with the focus of the event. The Bucket List reference can simply be an acknowledgement of the peripheral ideas and concepts that seemed worth exploring in more detail at a later date. Some of these might be related to other operations or processes that could benefit from ideas developed during the event. Others might be peripheral issues related to the product or process that there was no time to develop during the event. Capturing these items allows the team to take one more look at them at the end of the 3P event week to decide if they should be pursued further in the days ahead. It brings closure to other great ideas that came out of the event.

The Action Items should cover the short- to medium-term needs for progressing the project. They should not be actions already incorporated in the overall project plan. There is no need for that redundancy. They should consider presentation of the 3P final prototype to other groups in the company. It might include key communications or future team meetings or vendor visits. Whatever the actions, they should all follow the pattern of having a clearly defined task, a due date for delivery, and one person identified to be responsible for accomplishing the action. Also consider who the action owner will be accountable to for follow up. This is a pretty common approach, but one that is not commonly followed. It is a good standard to develop when using the 3P process.

Final Report-Out Session

With all of the above elements prepared in as much detail as time allows, the report-out meeting will go fairly quickly. Although all of the participants have been involved in putting together the necessary information, most of them will not be aware of the other groups' efforts, so everyone will learn during the report-out step. The Event Judges will attend, and sometimes other senior managers are invited also, but choose carefully when inviting other outsiders if the changes being presented will need significant effort to enable them to understand.

The final walk-through of the combined prototype will ground everyone in the new methods being employed. Often the team will demonstrate how normal operations will work and possibly cover product changeovers during this part of the presentation. Referring to the Process at a Glance during that presentation is powerful for enabling understanding and demonstrating the depth of detail that was considered. Ranking each functional step using the evaluation criteria will further increase understanding and appreciation of the benefits of the new design.

The goals presentation and summary bring it all together and highlight the success of the event. This is usually followed up with the preliminary project plan and a request for use of resources from other departments in order to move the new line forward toward an on-time and on-budget completion. This discussion can be followed up with the action plans and recognition of any bucket list items that might be relevant.

Hopefully the team has taken many pictures of the work accomplished during the week. As noted, because the three prototypes get taken apart and reconfigured, having pictures serves as a great reminder of the thought process used throughout the process.

Every event should recognize the hard work and amazing ideas generated in such a short amount of time. 3P brings out the best in a cross-functional team, enabling them to contribute as individuals, applying their own innovation and creativity, as well as leveraging other's ideas in the group and building upon them to make them even better. Lean 3P encourages constructive conflict between functional groups and helps them optimize their collective needs. Finally it empowers the team with management's understanding and support to try new things, sharing risk together, so that the company will reap the gains. 3P enables good people to accomplish great things never thought possible with other methods of design.

Now let us take one more visit to Kathryn's Finest's 3P event to see how they have applied all of these principles to their Baked Apple Project.

Example of Developing the Final Prototype—Baked Apple Project

When Kathryn's Finest participants returned the next day after reporting-out on each of their three prototypes, the mood was bittersweet. They decided to document the progress they made by having each team stand in front of their prototype and getting a group shot, allowing a few minutes to consider the innovation they progressed in the past couple of days. Each team took pictures of the different alternatives they built into the prototype and also took pictures of their Process at a Glance and the scoring they had completed during their evaluation stage. This was done both for nostalgia and celebratory purposes but more so to capture the key learnings. There was a chance that one or more of the intended final designs would need to be modified or even abandoned as the design and build stages continued. Having well-documented alternate approaches could provide a ready path for any future changes required.

The Facilitators called the group together and briefly described the requirements of the final report-out session. The Vice President of Operations, Alexandra, stepped in and added to what the Facilitators had just presented and suggested that they split up into three new groups in order to get all of the required steps completed effectively, utilizing the strengths of all of the Process Experts. She suggested the following breakdown of the participants and appointed leads for each of the teams.

PROTOTYPE BUILD TEAM

Responsible for

- ☑ Full construction of the final prototype
- ☑ Layout of the operation with material and tool storage
- ☑ Changeover and sanitization plan
- ☑ Takt time calculations
- ☑ Spaghetti diagram (product flow, people flow)
- ☑ Final Process at a Glance

People assigned were as follows:

- ☑ Production Supervisor—Tuan (Team Lead)
- ☑ Process Support Engineers—Al R, Bill M, Alan M
- ☑ Design Engineers—Philip, Carl J
- ☑ Operators—Jimmy B, Stuart
- ☑ Material Handler—Gary A
- ☑ Receiving/Shipping—Mike S
- ☑ Facilities—Earl, Glenn
- ☑ Quality Control—John D

The Prototype Build Team will utilize the skills of the Operations people, Engineers, Facilities, and Logistics people to physically build the final prototype, considering all of the evaluation criteria and the needs of each of the stakeholders on the team. The quality control person was also a key member because the operation needed to be sanitary, employing good manufacturing practices (GMP) throughout the process, and quality controlled to ensure proper flavor and appearance of the final product.

FINANCIAL COST ESTIMATE TEAM

Responsible for

- ☑ Creating final cost estimate
- ☑ Spend map for cash flow
- ☑ Developing preliminary bidding structure for new equipment and services
- ☑ Identifying key suppliers and initiating quotations, procurement protocols

People assigned were as follows:

- ☑ Finance—Rose C (Team Lead)
- ☑ Procurement—Nancy Z
- ☑ Senior Design Engineer—Matt D

PROJECT PLANNING TEAM

This team is responsible for

- ☑ Resource plan and teams
- ☑ Project timeline with milestones
- ☑ Process at a Glance contributions
- ☑ Proposed design input requirements (DIR)

People assigned were as follows:

- ☑ Vice President Operations—Alexandra C (Team Lead)
- ☑ New Product Development Manager—Charles F
- ☑ Director of Marketing and Sales—Jon G
- ☑ Operations Manager—Paul G
- ☑ Chief Baker—Loretta S

After the assignments were handed out, each team met separately to develop a preliminary plan of attack. Time was very limited and there was a great amount of work to be done.

The Prototype Build Team decided to construct a new Process at a Glance for the final process. They would add detail to it over the hours ahead, but it would serve as a roadmap to help them in the construction. Over the months ahead this same Process at a Glance would stay with the project and be updated as the development of the design concepts continued through to final delivery of the operation. The team also agreed to include the key transition steps in their Process at a Glance along with a function-by-function cost analysis and takt time. This was decided to ensure visibility for the money being spent on each section and to ensure the overall takt time requirement was met at every step within the process. Overbuilding would add needless cost. Underbuilding might create a loss of capacity.

The Financial Cost Estimate Team laid out a plan, identifying the key vendors they would contact. Negotiations were continuing with the Farmer Brown Company to reduce the cost of their apples, and some progress was being made. A separate team was dispatched to visit their location and to understand the upcharge they were requesting for the tighter-specification apples. They would also work with the other two teams to develop allocation of resource costs, facilities costs, and other necessary expenses, while creating a high-level spend map in order to determine cash flow requirements.

Just as they were about to start working with the key equipment vendors to estimate the line costs, Earl, the event Facilities representative, walked over with a grim look on his face. They had received three quotes on a new food-grade clean room and the costs were double the original estimates. Jaws dropped. The clean-room costs were already a big part of the project cost. This was overwhelming and with three close quotes, it was unlikely that they could negotiate their way out of the problem.

Word of the clean-room issue spread like wildfire. Even if they got the apple price fixed, they could never cut enough costs from the equipment to compensate for this facility increase. This time it was Paul, the Operations Manager, who stepped up to calm the group. He offered to meet with Glenn and Earl from Facilities and see if there were any options for reducing the cost. Perhaps they could reduce their standards or decrease the space foot-print? Paul encouraged the group to continue working while Earl and he tried to tackle this latest problem.

The Project Planning Team initially spent their time with the Prototype Team, trying to understand the scope of development of each of the functional aspects of the new operation. How much discovery would be required? What were the greatest risks technically? What other risks needed to be considered? People, resources, availability of the build space, and stor-age for the incoming equipment prior to build could be major factors. What if customer test panels thought that the taste of the "new and improved baked apples" was too different from the traditional home-baked flavor? Could they accept that risk or would they have to go back to the original method of baking? They started developing probability versus risk models and would build specific milestones into the project plan to address them. For the most significant and most probable risks, they discussed developing both the high-risk and a safe alternative concurrently to avoid any problem that would derail their project timeline too late in the process to compensate and still hit their delivery targets. This decision would cost additional money and resources and there would likely be a write-off of associated expenses if executed, but they recognized that the cost of that would be less if done early in the project.

■ Section Summary

☑ Optimization Phase Planning Session
☑ Distribution of Responsibilities

FIRST LOOK AT THE FINAL PROTOTYPE PROCESS—BAKED APPLE PROJECT

Working together, the Prototype Build Team and the Project Planning Team developed their combined Process at a Glance for the final prototype. While they developed this, a great deal of discussion occurred that shaped the way the new operation would be built. To capture the many changes, a few people worked on a sketch of the combined operation. Figure 12.1 represents their new process.

The Prototype Build Team began developing the final prototype. In the following section we'll follow along with the types of analysis that they did while focusing on the specific value-adding process steps and transitions.

ACCUMULATION

Apples would be supplied to the operation in large bulk bags that held 1000 pounds of apples. They would be received daily from Farmer Brown,

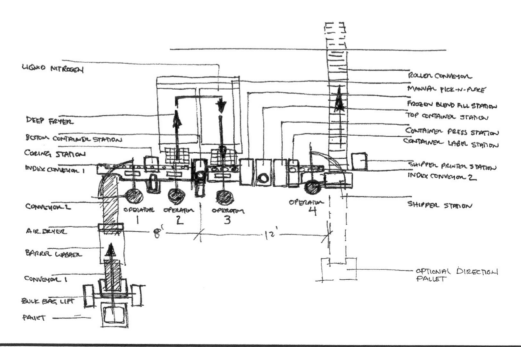

Figure 12.1   Sketch of the final prototype layout.

the supplier, on pallets, and be moved from the receiving area into a small cold room for day storage while quality inspections were conducted. When qualified and Operations were ready for them, the apples would be taken to the operating area by a lift truck and lifted over top of the unloading conveyor using an overhead hoist system. When in place a tie was loosened on the bottom of the bag to allow the apples to gradually flow out and onto a rubberized moving conveyor. The bulk bag system and conveyor were both depicted on the Process at a Glance representing the function *and* the transition step. One bulk bag held 1000 pounds of apples, and at an average weight of 8 oz per apple, this would calculate to about 2000 apples. With a required takt time of 31 apples per minute, a bulk bag would provide about 1 hour of run time. The team agreed this was acceptable but made a note to think about how quickly and easily the bags could be changed out and replaced with a full bag. Some buffer might be needed. Figure 12.2 shows the bulk-bag unloading system in operation.

☑   Final Prototype Analysis

The team deliberated the next function. Originally they had designed a step for separating the bad apples from the good. Because the quality of the apples was very high and the anticipated number of rejects very low, they agreed it would be better to go directly to the washing step and separate bad apples afterward.

Figure 12.2   Apple bulk bag loading in operation.

WASH

The conveyor fed the apples into a commercial barrel washer, a batch operation that completely submerged the apples in swirling water that thoroughly cleansed them, ejecting the dirty water, and using purified water for the final rinse step. The process took 7 minutes to complete and could wash a maximum of 500 apples per batch. Takt time for the barrel washer was deemed acceptable. Figure 12.3 shows the apples being conveyed into the prototype barrel washer.

DRY

The washed apples exiting the washer were immediately passed under an air-knife system that used a high-pressure stream of purified air to blow off the water droplets. The cost of the air-knife system was relatively low, and the benefits of immediately drying the apples would eliminate any potential microbial concerns associated with leaving wet apples on a conveyor belt. Even though the apples would be heated in a subsequent step, quality was paramount, and keeping the apples clean and free of any microbial attack was important. They made a note to auto-unload the barrel washer.

REMOVE MATERIAL

Exiting the drying system, the apples would move on a conveyor belt and be presented in front of an operator. At the operator station they decided to have the person pick up each apple and load it into a fixture upside down, in the

Figure 12.3   The prototype barrel washing in operation.

correct position for the next functional step, cutting out the apple core. In the process of handling each apple, the operator could do an inspection to look for bruising or damage that might affect product quality. Bad apples would be discarded in a recycle bin adjacent to the station, and good apples would be placed in position in a nest. To ensure continuous flow, the operators would rotate off on a periodic basis. This would keep the line going during breaks and prevent any ergonomic repetitive motion stress concerns from developing.

The fixtured apple would now be indexed the apple to the next position where a top support would drop down at the same time a rotary knife would come up from underneath the machine, cutting and also scooping out the apple core and dropping it into a waste bin. They noted that this was not a commercially available method, but research had found examples of this application that they felt could be successfully developed. They made a note on the Process at a Glance chart that this might have an elevated risk because it was not proven in this application. A fallback alternative might be necessary. One team member also noted that access underneath must be available to ensure the ability to thoroughly clean the equipment.

HEAT

With the plan to deep fry the apples in hot butter and sugar, the step to fill the apple would not be needed yet. Instead, the apples were indexed to another operator station where a person would remove it, do an inspection, and place them 24 at a time, into a basket-type device made of stainless steel and loaded onto a type of trolley indexing system. The machine would have two fryers, and the baskets would be inserted for a specified amount of cooking time. Only one fryer would be active at any one time allowing the other to be changed out with fresh butter without interrupting flow. Details of the cleaning procedure would be needed, and this was noted on the chart. The premixed butter and brown sugar would come from a supplier they had contacted in package sizes that matched the exact requirements of the fryer. Sprinklers and other safety considerations would also need to be incorporated into the design.

COOL

After the cooking process was complete, the baskets would be lifted out of the fryer by a trolley system and moved into another chamber where the apples would be flash frozen by spraying liquid nitrogen on them. This was a large commercial module that needed some customization but would meet the requirements and provide all of the safety interlocks necessary for doing this potentially hazardous step. The team noted that this was one of the most expensive aspects of the new operation, but the benefits of stopping the cooking process quickly and protecting the apple from degradation by freezing seemed worth the high price. The team's prototype of the heating and flash-freezing modules can be seen in Figure 12.4.

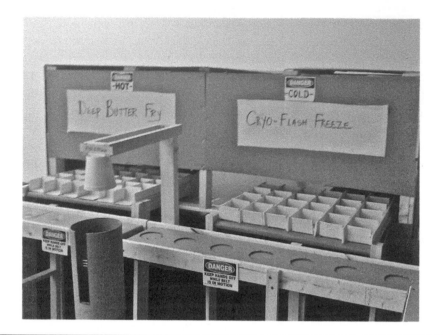

Figure 12.4 Apple "deep fryer" and flash-freeze prototypes.

PROTECT

After exiting the flash cooler, the apples, still in the baskets, were picked up via a manually operated vacuum assist pick and place unit and placed into a translucent clamshell package. The clamshells would be dispensed from an automatic dispensing machine designed for that purpose and placed on a conveyor adjacent to the frozen apple pick and place. Having an operator involved in this step allowed for additional inspection and assured that the apples were placed correctly in the package.

FILL

The apples, now located in the clamshell, were then moved under another piece of equipment that placed a frozen mix of brown sugar and butter, perfectly shaped to match the hole, into the apple. This machine was a modified pick and place unit that would require customization. The supplier of the sugary insert agreed to work with the designer of the pick and place and provide the inserts in a type of "sleeve" that allowed the pick and place to grab them individually. The Operations Manager made a comment and added a note to the Process at a Glance board that this approach might prove to be too complex to operate reliably and the alternative could be another manual step if required. The engineers balked at his conservative comments, believing that they could make it work, but his years of experience threw up a caution sign. Some early prototyping of this concept would need to be done to test the reliability before making a big investment in the

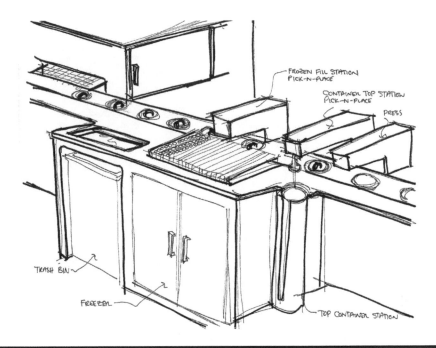

Figure 12.5   Core filling operation using frozen butter/sugar inserts.

technology. In addition, the finance team would do another review of the cost–benefit relationship of the two approaches. The designer's sketch of the apple core filling operation can be seen in Figure 12.5, followed by a photo of the prototype in Figure 12.6.

IDENTIFY

The clamshell would continue down the conveyor and a guide system closed it up while another top plate would come down and snap the lid closed. From there they would continue under a print and apply labeling system, where preprinted label stock would have a lot number and use-by date printed on it before being applied to the top of the clamshell. The print and apply system was a fairly standard labeling system, commercially available.

GROUP

The last stop for the operation was final packaging into corrugated shipping cartons that an operator would construct, tape, and fill with 24 clamshells. Another printer was sitting nearby on a tabletop, ready to make up the shipper labels that the operator would apply once the carton was ready for shipment. The filled cartons were on a gravity conveyor with rollers, where they would flow directly into the cold room, where they would be put away and into inventory by a material handler. They considered another option for the finished apples to come out onto a pallet in the operating area. There

Figure 12.6   Prototype of the apple core filling operation.

was a time limit on how long the frozen apples were able to stay out of the cold room, and an andon light system was planned to warn when the time limit was being approached. A small refrigerator next to the packaging station was available for quality samples to be placed. The quality department personnel could come out to pick up the samples for testing. It also gave them the opportunity to see what was happening on the line. Figure 12.7 shows the finished product, labeled and packed in the shipping container.

Figure 12.7   The finished product in the shipping container.

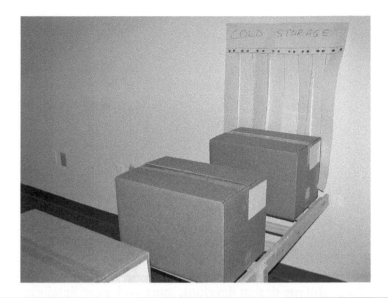

Figure 12.8   The gravity conveyor, showing product moving into the adjacent cold storage room.

In Figure 12.8 the packaged product is fed into an adjacent cold room. The attention to detail that is built into the final prototype can be seen in Figures 12.9 and 12.10, where preventative maintenance and a material supermarket are shown, respectively.

Example of the 3P Final Report-Out Session—Baked Apple Project

For two days the teams worked diligently. When the time for the report-out session came, there was a collective sigh of relief by the teams. In the final hours each team had been frantically attempting to complete their respective assignments. Recognizing the impossibility of the challenge forced them to prioritize their activities and make calculated assumptions that they noted to follow up on later. The Judges had returned to the meeting place and had wandered around the final prototype amazed at the progress that had been made.

PROCESS AT A GLANCE

The Prototype Build Team went first and had everyone gather around a very large Process at a Glance chart that can be seen in Table 12.1. It was much larger than the previous versions of the chart used for the initial three prototypes. Added to the successive functional steps were each of the transition steps. In addition, they included a row for the estimated cost of each step and another for takt time. Glancing across the large chart quickly showed where the money was going and also verified that every process and transition step should be capable of meeting the required takt time

Figure 12.9 **As the prototype design develops, more and more details are added. This shows preventative maintenance detail.**

Figure 12.10 **Prototype details include "supermarket" storage for required supplies.**

rates. A couple of the functions could be questioned to see if they might be overbuilt because they exceeded takt by such a great amount. Overall everything looked very reasonable.

The team leader walked the group through the process function by function, indicating where they had concerns for technical risk and other aspects of the proposed design that had been considered.

Underneath the Process at a Glance (Table 12.1), the evaluation criteria had been listed (Table 12.2), and every functional step had been rated. Not every aspect of the transition steps had been filled out, but it was a very comprehensive evaluation. There was good discussion during the presentation, and a couple of the ratings were modified based on input from the other participants and the Judges. In every case the designs were deemed better than or equivalent to the traditional design, with the majority considered significantly better. On a separate sheet they evaluated and rated the overall design (Table 12.3).

Takt time was calculated as a minimum, maximum, and average value, discounting the two highest values to give a more reasonable number. The two high numbers were related to the receipt of the bulk bags of apples, and although they were arguably overbuilt, the cost penalty versus the benefit of handling them this way made the exception acceptable.

The rolled-up cost of the equipment was estimated, but it was noted that this was not a final project cost estimate, just relative estimates of what each element of the operation would cost, mostly based on the equipment costs. Installation, facilities work, engineering time, and contingencies would be estimated in greater detail later. It was valuable to have costs included on the Process at a Glance to see relative costs of each step and for the overall process.

Table 12.1 depicts the final Process at a Glance diagram that was completed for the new line. The team filled out as much of the information as they could, recognizing that the work would continue after the event, and additional detail would continue to be added.

■ Section Summary

☑ Sketch the layout of the final prototype
☑ Build the final prototype
☑ Add features and details
☑ Develop the final Process at a Glance

FINAL PROTOTYPE WALK-THROUGH

Two production operators, Jimmy and Stuart, led the group through the presentation of the final prototype. Their enthusiasm was contagious as they went from step to step in the process, pointing out operational elements of which they were especially proud. Locating tools close to their intended point

Table 12.1 Kathryn's Finest Lean 3P Final Process-at-a-Glance

Process step	Transition step	Material	Method	Gauge	Tool	Fixture or jig	Machine	Takt time	Cost $ (000)
(transition)	Load bulk bags		Overhead hoist system	Mark floor for apple bulk bag	Height adjustment bar	None	Safety rails to block hoist path	300 apm	$25
1 Accumulate		Bulk bags	Bag dumper and filler	None	None	None	None	75 apm	$10
(transition)	Move to washer	Apples	Conveyor or belt	None	None	Start-stop button	None	35 apm	$5
2 Wash			Barrel washer (two units)	None	Barrel clearing basket	None	Guarding and purified water	40 apm	$170
3 Dry		Clean/dry apples	Air-knife system (two units)	None	None	Yes	Noise insulation and filtered air	40 apm	$50
(transition)	Move to operator station	Convey-or or belt	Convey-or or belt	None	None	None	e-Stop button	35 apm	$5
4 Separate		Operator place in fixture	Manual separation and placement in nest	Human inspection	Sizing tool	Recycle bin for rejected apples	Yes	35 apm	$2
5 Make hole		Knife cuts	Rotary knife with scoop	Limit switch on knife	Yes—top plate support	e-Stop button	Easy access to clean/maintain	32 apm	$45
(transition)	Move to frying basket	Operator place in basket	Manual placement of apples in basket	Human inspection	Magnifying glass	Fry basket	Fry basket	35 apm	$5
6 Heat		Deep fry	Custom deep-frying equipment	Temperature monitor and control	Temperature gauge for verification	Fry basket	Guarding and clean-out system	40 apm	$100
7 Cool		Flash cool	Liquid N2 flash cooling	Temperature monitor and control	Cryogenic apple grabber	Fry basket	Oxygen sensor safety system	40 apm	$40
(transition)	Move from flash cool to packing		Trolley lifts and indexes baskets	None	None	Fry basket	None	33 apm	$65
8 Protect		Clear clamshell	Clamshell packing system	None	None	Indexing conveyor	None	35 apm	$40
(transition)	Load into clamshell container	Operator using vacuum lift	Vacuum assist placement	Vacuum gauge	Manual operated suction lift	None	None	32 apm	$5
9 Fill		Pick and place	Pick and place frozen blend	Vacuum gauge	Manual operated suction lift	Frozen butter cartridge	e-Stop button	34 apm	$15
(transition)	Close and seal clamshell		Conveyor with closure guide; Automatic folder and top-seal	Limit switch	None	Guarding	Yes	32 apm	$15
10 Identify		Printed label	Print and apply system	Bar code scanner	None	Labeling system access to maintain	Yes	40 apm	$60
(transition)	Auto-unload to packaging	Conveyor with collection buffer	Conveyor or belt	Line clear sensor	None	None	None	40 apm	$5
11 Group		Manual packaging	Corrugated shipping carton	Mark floor for pallet positions	Manual operated suction lift	None	Off-line shipper label machine	32 apm	$15

Table 12.2 Kathryn's Finest Lean 3P Process-at-a-Glance with Criteria Evaluation by Module

Process step	1 Accumulate		2 Wash	3 Dry		4 Separate	5 Make hole		6 Heat	7 Cool		8 Protect		9 Fill		10 Identify		11 Group
Transition step	Load bulk bags	Move to washer			Move to operator station			Move to frying basket			Move-flash cool to packing		Load into clamshell container		Close and seal clamshell		Auto-unload to packaging	
Material	Bulk bags	Apples	Sub-merse water wash	Forced air	Clean/dry apples	Operator place in fixture	Knife cuts	Operator place in basket	Deep fry	Flash cool	Basket on trolly system	Clear clamshell	Operator using vacuum lift	Pick and place	Conveyor with closure guide	Printed label	Conveyor with collection	Corrugated shipping carton
Method	Overhead hoist; Bag dumper/filler	Conveyor belt	Barrel washer	Air-knife system	Conveyor belt	Manual placement	Rotary knife	Manual placement	Deep-fry equipment	Liquid N2 cooling	Trolley system	Clamshell packing	Vacuum placement	Pick/place frozen blend	Automatic fold/seal	Print and apply	Convey-belt	Manual packaging

Evaluation Criteria

Takt Time	75	35	40	40	35	35	32	35	40	40	33	35	32	34	32	40	40	32
Minimal Capital	0	−	0	0	0	0	0	0	−	0	0	0	0	0	0	0	0	0
Pull System	0	+	0	+	0	0	0	0	0	0	0	0	0	+	0	+	0	0
People Involvement	+	0	0	+	0	+	0	+	+	+	+	+	+	+	+	+	+	+
Automatic Unloading	+	0	+	+	0	0	0	0	0	0	0	0	0	0	0	0	0	0
Load-Load Operations	0	+	0	0	+	0	0	+	0	0	0	0	0	+	0	+	0	0
Mistake Proof (Poka-Yoke)	+	0	+	0	0	+	+	0	0	0	0	+	+	+	+	+	0	+
Minimal Space Required	0	+	+	+	+	+	0	+	0	0	0	+	0	0	0	0	0	0
100% Gauging	0	+	0	0	0	+	0	+	0	0	0	0	0	+	0	+	0	0
Safety, Ergonomics	0	0	0	0	0	0	0	0	0	0	0	0	+	+	+	+	+	+
Internal Waste Collection	0	0	0	0	0	+	−	0	0	0	0	0	0	0	0	0	0	0
Simple as Possible	+	0	0	+	+	+	+	+	+	+	0	+	0	+	0	0	0	+
Standard/Off-the-Shelf	+	+	+	0	0	+	0	0	0	0	0	+	+	0	+	+	+	+
Process Capability	+	0	0	0	0	0	−	0	+	+	0	0	0	0	0	0	0	0
Maintenance Free	0	0	−	+	0	+	0	+	+	+	+	+	0	0	+	+	+	+
Autonomation	0	+	+	0	0	0	0	0	+	+	0	0	0	+	0	+	+	+
Scalability	0	0	0	+	0	+	0	+	0	0	+	0	0	0	0	+	+	+
TOTAL +	5	2	6	7	4	9	4	5	5	5	3	4	2	7	3	7	4	3
TOTAL −	2	0	2	0	0	0	0	0	1	1	0	0	0	0	0	0	0	0
RATING TOTAL	3+	2+	4+	7+	4+	9+	4+	5+	4+	4+	3+	4+	2+	7+	3+	7+	4+	3+

Table 12.3 Kathryn's Finest Lean 3P Overall Criteria Evaluation

Evaluation Criteria	Weighting	Score	Comments
Takt Time	2	+ +	Very balanced
Minimal Capital	2	+ +	Two lines - half initial CAPEX
Pull System		+	
People Involvement		+	
Automatic Unloading		0	
Load-Load Operations		0	
Mistake Proof (Poka-Yoke)		0	
Minimal Space Required		+	
100% Gauging		0	
Safety, Ergonomics	2	0	
Internal Waste Collection		0	
Simple as Possible	2	+ +	Good balance of people and automation
Standard/Off-the-Shelf		+	
Process Capability		0	
Maintenance Free		+	
Autonomation		+	
Scalability		0	Build two lines
TOTAL +		12 +	
TOTAL −		none	
RATING TOTAL		12 +	EXCELLENT

of use and side panel doors that provided easy access to clean or perform maintenance were pointed out. The layout of the overall operation would enable operators to work close by each other. This would enhance communication but would also enable them to help each other if someone was falling behind or had a problem.

Throughout the walk-through questions came up and were fielded by one of the team members. The maintenance people and engineers had many points of reference for how they would build in reliability. Grease fitting manifolds to enable lubrication, removable panels for easy access, and even minor details like the location for keeping replacement high-wear parts had been considered and depicted on the model.

Takt time was met throughout each module and for the overall operation. Some pieces of equipment would require testing to ensure that the estimated takt time could be achieved, and those were noted as higher risk. They had communicated this to the team working on the project timeline and resource plan to ensure that they address the risk early.

Perhaps the most impressive feature of the operation was the way they had designed in the operator's role. In every step that required intricate handing or an important quality inspection, they had inserted a person versus a highly complex automated solution. This saved considerable initial costs but also reduced complexity and risk. People had typically grown up coring apples and inspecting for left-behind pieces of core. People were also very good at positioning apples in the proper orientation in a nest. They recognized that machines and vision systems would struggle with getting it right consistently, considering the irregular shape of the apples. Natural variation in any material is a common problem for manufacturing sysems, often requiring significant technical adjustments before eventually working correctly. People tend to get it right almost immediately when properly trained. Placing people in this role also ensured their engagement and ownership of the process and the product. The operators suggested a number of mistake-proofing features that would enable them to be consistently successful. They understood better than anyone the places where there was the greatest risk of error and helped put countermeasures in place to prevent mistakes from occurring.

Although there were no plans for changing over to different products on this new operation, there would be considerable need to ensure cleanliness and sanitization. The team had thought through the details and had access panels for cleaning, locations for sanitization solutions, and wash rags. They walked through each part of the sanitization process in detail, even including a changeover time analysis in their estimates. It was amazing. Normally this type of activity was done after the line was nearly complete and was done by the Operations team with little involvement from engineering. Using the Lean 3P approach everyone had considered the requirements and the designs were improved to make changeovers quick and easy.

The strong emphasis on mistake-proofing and ensuring a smooth flow of operators working in the process also correlated to improved product and materials flow into and out of the area. The operators and material handling group had a vested interest in making it work smoothly, and the physical model allowed them to see, develop, and ultimately present the benefits to the larger team of people.

SPAGHETTI DIAGRAM

The final evaluation criteria that the Prototype Build Team was responsible for was presentation of the new Spaghetti diagram. They had used the prototype to walk through how the operators would interact with the materials and the process equipment during normal operation and during changeover and sanitization. They tracked the flow of the raw materials and the finished goods to ensure one-pass flow wherever possible. They even analyzed the locations of the tools that would be required and the supplies that they would need, and built that into the prototype model. When they compared the original estimated Spaghetti diagram to the one they had created based on the new process model, the results were staggering. They had fully achieved a 75% reduction in the people-movement and material distances traveled using the new process.

Everyone was duly impressed when the team concluded their presentation. They had worked very hard and delivered a very well thought-out process design for their new product.

- ■ Section Summary
  - ☑ Present the final prototype
  - ☑ Demonstrate operator functions
  - ☑ Demonstrate quick-changeover approach
  - ☑ Point out details included for material handling, maintenance, and material storage
  - ☑ Spaghetti Diagram

FINAL COST ESTIMATE TEAM

The next team to present was the Final Cost Estimate Team. Throughout the past day they worked together with the Prototype Build Team to get details about the equipment they planned to use and specific suppliers they planned to use. Although Kathryn's Finest had a procurement policy that required competitive bids whenever possible, they also respected equipment standardization and recognized the need for sole source quotes for very specialized equipment. After identifying the major equipment components, Nancy from Procurement began contacting potential vendors and began laying the foundation for company terms and conditions of sale and their bidding process. In several instances she was able to get budgetary quotes from the vendors in order to complete a project cost estimate. Matt from Design Engineering was involved in a number of conference calls with the vendors, answering technical questions and evaluating proposed alternatives.

Concurrent with the equipment procurement activities under way, Rose, from Finance, and Matt had also been busy working with the Project Planning Team to develop costs for the engineering time that could be capitalized along with the labor costs to physically build the new operation. They worked closely with the planning team who were focused on understanding the complexity of the Design and Build process to determine timing and resource requirements. They also reviewed the required facilities' alterations, estimating some minor structural changes, heating, ventilating, air conditioning (HVAC) improvements, lighting upgrades, and permitting requirements.

The Final Cost Estimate Team looked at the resources needed and the number of people who would be full time on the project. Many others, however, would only be working part time on it in order to keep their normal jobs going. This was the most challenging aspect of the planning process. How could they account for resource time for people who had multiple responsibilities? Everyone recognized the need to have the project operational in time for a Fall launch, leaving only 9 months to complete a huge effort. The speed of the project would almost certainly require extra expense for overtime and additional outside resources. The Project Planning Team resolved that this was one of the greatest risks to the project timeline and budget. They would meet daily and weekly with the team, using a visual process to track when their milestones were approaching.

Two big hurdles still faced the team financially. Farmer Brown's high premium for the select-grade apples, and the excessive clean room costs. Either of these could kill the project.

The team who visited Farmer Brown learned why the cost had gone up so much. Their specification for apple size had required a lot of special handling and segregation, even though it was in the range of the most common-sized apples available. In addition, Kathryn's Finest had very strict terms and conditions (T&C's) for suppliers, with significant penalties for supply of out of specification product. When the team returned to the event with this news, the Prototype Build team went to work evaluating the amount of variation they could actually tolerate. It turned out to be much larger than specified. Nancy from procurement also got involved and had worked with Kathryn's legal group to see if the terms and conditions could be modified. This turned out to be a brick wall. T&C's would not be changed. Facing a loss of new business, Farmer Brown took a look at the new, broader specifications and finally agreed to them with only a 5% upcharge over the commodity price. This still had an impact on final product cost, but it was now a viable product again.

The second hurdle for the clean room was not going to be negotiated away. The special team investigating this issue came back with a new proposal for the event team. Glenn the Facilities Manager had taken the challenge to reduce the cost of the new room. Paul, Earl, and he had called vendors and sister plants for alternatives, even considering used equipment as an option. Food-grade clean room standards were necessary, and lowering the quality specifications was not an option. What if they could build it themselves? A lot of the cost of the commercial rooms was in the prefabricated structure. The Facilities organization had a lot of experience building out office complexes and conference rooms. Why not a clean room? They certainly had the experience with the clean room's HVAC requirements because they routinely maintained them. They often installed the special air filtration systems required and did similar flooring projects using easy-to-clean epoxy floors. They actually had extra high-efficiency particulate air (HEPA) filters and special lighting available from another project that they determined to meet the requirements of the new room.

The Facilities team put together a plan and an estimate for creating their own clean room. The cost was half of the original cost, and one quarter that of the quotations. All of a sudden, the project financials went from being highly unfavorable to favorable again.

With an understanding of the remaining risks to the project well defined, the Final Cost Estimate Team had built financial contingencies into the areas of greatest risk based on key input from the Project Planning Team. When finished they had a very good estimate for the equipment, facilities, and resource requirements for the project, along with a spend map they developed with the Project Planning Team for estimating cash flow. Part of their estimate considered the disruption of normal operations during some periods of the construction process and the need to temporarily build up inventory levels of other products.

The Final Cost Estimate Team's report-out session was well-received. Comparing the costs to the original estimates developed during the phase-gate process, the capital cost of the new line was about 25% lower, thanks to the excellent work of the Facilities goup. The ongoing operational costs were also favorable in spite of the 5% increase in the apple supply costs. They estimated that total labor costs would be 20% lower and inventory levels would be nearly half of their previous estimates. The entire team contributed to significant value creation through their innovative design and creativity.

- ■ Section Summary

    - ☑ Present high-level cost estimate
    - ☑ Present cash flow timeline
    - ☑ Identify known risks and mitigation methods
    - ☑ Compare financial estimates to event targets

PROJECT PLANNING TEAM

The last group to present was the Project Planning Team. They had interacted extensively with the two other teams developing the plan estimates. The Project Planning Team needed to understand all facets of the project. They had to know equipment lead times, the scope of facility

Figure 12.11   The report-out session, and a review of the project and financial plans.

modifications, and the probable impact on routine operations in order
to estimate potential inventory builds and overtime requirements during
the construction phase. Figure 12.11 depicts the Project Planning team's
report-out session.

Their most critical task was identifying the areas of greatest risk to the project. Risk came from many areas. Technical risk from application of new technologies was very real and required time and resources to understand and
mitigate. Resource risk was a major issue for them given the tight timeline and
the number of people who would be shared with other projects. Laboriously
they developed the scenarios and developed key milestones, working back-
wards from the targeted launch date till the official start of the project. At this
time the start date was still uncertain as the "Approval for Capital Investment"
request had not been developed for management approval.

Risks were identified, and contingencies and buffers were built into the
plans for achieving the requirement of a launch 9 months from the event.
Added costs to mitigate the risks were built in proportional to the probability of requiring it. They recognized that they could not build in costs to fully
fund every contingency and still have a viable project.

The team sketched out a project plan and identified teams of people
who would be responsible for key aspects of the development. They
showed hand-offs from one group to another and noted those areas as
high risk because special attention was required to ensure a good transition without losing time. The Project Planning Team worked to develop
definitions of tasks to provide guidance to the group responsible for

delivering it. This was important to ensure that the design deliverables were met, but it also ensured that they did not needlessly overdesign any aspects of the operation. The milestones considered input regarding risk, from the people involved in each task. Overall, lack of resources was the greatest risk. They got commitments from the department leaders to prioritize the availability of people for this key project. A project "war" room would be set up and progress boards used to track performance, monitor mitigation of risk, and serve as a meeting place for the team. They would have a short stand-up meeting daily for the first 2 weeks and evaluate a weekly meeting schedule as the project progressed. With the ability to keep the final prototype intact in the existing space, they agreed to keep the modifications current, adding detail and changes as often as they occurred. They would continue to use development of alternatives and prototyping when changes were necessary, and would always include the whole team.

Throughout the process the Project Planning Team interacted with the Final Cost Estimate Team and the Prototype Build Team. It was convenient that all three teams were co-located, facilitating excellent communication and enhancing progress at an unprecedented rate. Alexandra, the VP of Operations, commented several times that the rate of decision making that they were seeing would take weeks using their standard meeting processes, e-mails, and traditional hierarchies and silos, versus the hours it was taking here.

One point of emphasis was the need for their proposed cross-functional teams to stay together after the event. While most of them would go back to their normal jobs, some of their time would be needed to work with the design engineering team to be involved with the changes and decision process as the project progressed. They understood that some of the plans would change as they were developed further, and the benefits could only be sustained if the collaborative input continued to be utilized. Most of the subteams agreed to meet weekly at first with the option of meeting more or less frequently as needed.

The Project Planning Team went over the project timeline, highlighting key tasks and milestones that were necessary to deliver the project on time. It was aggressive, but everyone agreed it was achievable.

Together they developed an initial set of design input requirements (DIRs) for the new operation, utilizing some of the original targets such as takt time and process utility metrics, yield expectations, and a handful of other criteria. After walking through the prototype and stepping through the operational and changeover steps, everyone had a reasonable confidence level for attaining those DIR goals. Alexandra emphasized that they were only tentative specifications now. Further refinement would need to occur as they developed more project detail in the weeks ahead.

The final element of their presentation was a summary of the high-level impact of their design. They had met all of the major goals and honored the critical boundary conditions for the project, starting with the process for loading the apples onto the line and ending with palletized boxes of baked apples prepared to go into cold storage. The space required was slightly smaller than the amount allocated, and they actually moved some functional steps into the small footprint, avoiding material travel and enabling product flow. The schedule they developed stipulated a delivery in time for the fall launch, and they noted that it was aggressive but possible. Also, by improving flow they had been able to reduce inventory levels and meet the freezer constraints that were originally a big concern due to the cost of adding more frozen warehouse space. The design had built in all of the requirements for Good Manufacturing Practices as required.

At the conclusion of the event, the VP of Operations went around the room, personally thanking each person for his or her involvement and contributions. They took another set of group photographs in front of the final prototype and the large Process at a Glance chart. They agreed to keep the prototype up and updated for as long as possible to allow others from the company to look at the proposed new operation and to help promote the benefits they experienced using the Lean 3P Design process. Their Process at a Glance chart would be taken to a team room where it would be refreshed as the design teams continued to refine the ideas, adding more detail and making changes as they developed.

Kathryn's Finest had many challenges ahead of them as they developed their new operation and launched the new baked apple product. The 3P Event had been highly successful, exceeding expectations. Now the challenge would be to continue the collaborative approach to developing and delivering the operation on time and on budget, avoiding the tendency to slip back into old ways of working. In some regards this was their greatest challenge, but they tasted the sweet fruit of structured innovation. They had top-level support for the risks they were taking, and they had a new process for making their project status visual, in a way that fostered collaboration and expedient decision making. They were ready.

■ Section Summary

☑ Detailed Project Plan
☑ Understand and Plan for Mitigating Risks
☑ Design Input Requirements
☑ Recognize the Participants' Contribution

Summary

We have completed the steps that encompass the Lean 3P Event to develop both product and process concurrently, with consideration and input from all of

Figure 12.12 Two-thirds of Kathryn's Finest—some of the best friends and colleagues anyone could ask for.

the people involved in the new product development and manufacturing processes. In the following chapters we uncover some potential pitfalls and opportunities. Figure 12.12 shows the team at the close of the Lean 3P Event.

Developing the Final Prototype, Incorporating Evaluation Criteria, and Measuring Effectiveness, Final Report-Out Recap

☑ Optimize and finalize the hybrid prototype
☑ Report-out session
    ☑ Process at a Glance
    ☑ Estimated performance versus evaluation criteria
    ☑ Prototype walk-through, highlighting people involvement
    ☑ Financial plan and estimates versus target costs
    ☑ Project timeline and risks
    ☑ Open actions and opportunities
☑ Summary of accomplishments

Chapter 13

3P Project Management: Potholes and Stumbling Blocks

3P Project Management

When the excitement of the 3P event wanes and the participants return to their normal routines, the next 3P challenge strikes. The Facilitators have gone home. The e-mails and other demands that people have been postponing in order to devote a week of time to the event now await them. What behaviors now have to change in order to progress the new 3P concepts and ideas to fruition in the final operation?

Communication and Involvement

Perpetuating the 3P benefits after the event requires a new type of project team that includes some degree of active involvement from the original participants. As mentioned, some companies maintain the original participants as full-time members of the ongoing project team. Most of the time the next steps are predominantly engineering intensive, and without possessing those skills, it is unnecessary to keep a full-time team made up of the other stakeholders fully occupied. So how do you keep them actively involved on a part-time basis? Again, there is no textbook answer, but routine communication and progress reviews are a must.

Use of visual tools greatly helps to keep people engaged in a more effective way than standard meetings with people sitting around a conference table. Stand-up meetings around the final prototype while updating it with modifications and changes works well if the space is available to keep it intact after the event. Note that this is also a very powerful means of informally sharing the new concepts with the rest of the organization. It allows them to visualize the ideas and changes and continues to build ownership among people who will interact with the new operation. Stand-up meetings are short, well planned and focused, and should be held at a frequency appropriate to the project needs. Typically this will be weekly to biweekly and will involve all of the event participants (stakeholders).

Teams should always keep the Process at a Glance updated, continually refining the concepts and adding to the details embodied in each section. This in itself becomes a significant communication tool. Updates to each functional step can be covered very quickly, using the Process at a Glance to follow along sequentially. This is not a substitute for a project plan, but it does keep track of all upgrades and changes and enables people to see how the changes impact the big picture of the full operation.

Visual project boards work very well to communicate project status and issues. They enable people to quickly glance at the board and see exactly what the status is in key areas of the project, where the problem areas are, and if the schedule and costs are still on track based on progress against the project milestones. Using red and green symbols graphically shows the status without lengthy explanations required. Activities marked red indicate an issue that is not working or is behind schedule. Visual project boards are team boards that are updated on an agreed schedule with the full team present, each person posting new commitments or actions that he or she has accepted ownership of, and removing the ones that have been completed. Green status requires no further discussion. Red demands an action plan to mitigate the undesired status.

A3 Problem Solving

Every project incurs problems. Knowing that a problem exists and understanding the impact of it is the first step toward resolution. The A3 process is excellent for documenting and solving problems. A3 refers to nothing more than the size of a standard size of paper, approximately 11 inches by 17 inches in size. There are numerous excellent sources of information available on how and when to use an A3. In John Shook's book *Managing to Learn: Using the A3 Management Process to Solve Problems, Gain Agreement,*

Mentor and Lead, he indicates that A3 is the only Lean tool that can be found in every department in Toyota. The reason is that there is no standard format for it, no structured methodology, only a belief that every problem, no matter how complex or challenging, should be able to be succinctly expressed on a single sheet of paper.

With the understanding that A3 can take almost any form and utilize almost any format, let's go through the steps using one format that generally works pretty well for solving problems. It is a very good approach to capturing problems that occur during the development of a new operation because it makes the problem visual and encourages a lot of discussion and dialogue while it is being developed through every section. A3 closes the loops on the entire problem from start to finish with good documentation of the thinking process used to resolve it.

A problem-solving A3 starts with a succinct definition of the problem. The Problem Statement simply states what the deviation from the norm is that needs to be addressed. The statement of the problem should be articulated as accurately as possible. Charles F. Kettering, a U.S. electrical engineer and inventor, is quoted saying that "a problem well stated is a problem half solved" (Locher, 2011, p. 132). It is interesting how often a problem is misstated as a symptom related to the actual problem. Many times a problem will need to be restated several times during development of an A3 as we drill deeper into our understanding of the symptoms. A3 is a very iterative process, often involving input from many people from different functions, a trait that it shares with 3P.

The Background section expresses why the deviation is a problem, how big it is, and what the impact is on the organization, our customers, or the project. Current Condition refers to the status right now. In most cases these sections are addressed by going to the workplace, the site of where the problem occurs, and observing and talking to people involved with it. Pictures, sketches, and graphs are all helpful in telling the story of what is happening in a comprehensive, digestible manner.

The Goal section articulates the improved state that needs to be achieved. The goal can be visually depicted using a graph or a chart in addition to words. After the goal is determined, the Analysis of the problem is done, many times using the simple five "Why's" method for getting to the root of the problem. Fishbone diagrams are also commonly used. For highly complex problems consider using Keptner-Tregoe Analysis or Design of Experiments. In conducting the analysis any of the quality tools can be used to determine the ultimate root cause or causes. Most people want to solve the problem before setting a goal, but A3 intentionally seeks to set the goal ahead of this, so that clarity in the objective is not muddled by the difficulty of the problem. This enables deeper thinking when developing plans to mitigate the problem.

All of the previously mentioned steps are typically on the left-hand side of the document. All of them are there to help understand the nature of the undesired condition.

The right-hand side starts with the countermeasures developed to mitigate the identified root cause or causes of the problem and attain the desired goal state. With countermeasures developed, the next step is a defined Action Plan, where *who, what,* and *when* are defined. Create a list of action items required to achieve the countermeasures and reach the intended future state. Always define what will be done, when it must be done, and who is responsible for getting it done. The "who" should always be an individual, and not a group, as a group is very difficult to hold accountable.

Often many countermeasures are developed and not all of them are translated into action plans immediately. The highest probability plans can be executed first and if the goal is satisfied, nothing more needs to be done. When the goal is not fully attained, another countermeasure can be pulled out and developed with another pass through Shewhart's (or Deming's) Plan, Do, Check, Act cycle. The Follow-up section generally completes the A3 and is used to determine and execute plans to collect the data that will ensure that the actions have been completed, results measured, and the desired goal state has been achieved.

This process may seem demanding and tedious, but it is a disciplined approach to resolving problems, which provides excellent documentation of the thinking process used to resolve it. Beyond that, A3 encourages collaboration among stakeholders, and visits to the workplace where the problem is occurring in order to better understand the real issues. It has us do all of our "homework" before jumping to a solution. A3 is a very natural companion when using the Lean 3P Design process.

New Wine in Old Wineskins

In most companies design of new operations and new projects has certain protocols and procedures that still need to be adhered to. Some aspects of the design process might be covered within the Quality System. Formal concept reviews and design reviews are pretty standard in most companies.

Failure mode effect analysis (FMEA), hazard studies (HAZOPS), and other mechanisms for effectively reducing risk when making change still need to be adhered to and addressed. Over time as 3P becomes more embedded in the culture of the organization, some of these processes might be modified or even eliminated in favor of a better substitute; however, in the interim it is important to progress the formal steps in the accepted manner. When most of the participants in these standard reviews have also been participants in the 3P event, as is often the case, they go very well because people understand the process fully and there is little discovery. There is, however, some risk of preferential bias because people have that common background. Having some outsiders and an impartial, experienced facilitator for these studies reduces the risk of that occurring.

Out-of-the-Gate Performance

When a Lean 3P team delivers a project it should very quickly reach standard performance levels as measured by standard quality rates and standard costs. How quickly this level of performance is attained could be industry specific, but the goal is days to weeks, not months to years. Every project has problems, and every new operation will need some debottlenecking or adjustment period to fine-tune the process before standard performance can be demonstrated and sustained.

3P design often gravitates toward simpler solutions rather than complex ones. Simple solutions tend to cost less and be more reliable. This benefits the ongoing performance and helps achieve expected performance levels in less time.

The Process at a Glance also helps because it forces consideration of all the key elements of equipment design: Material, Method, Gauge, Tool, Fixture or Jig, and Machine considerations. If used correctly the Process at a Glance model can help avoid many problems that will impact performance, including early determination of normal material variation that must be accommodated by the operational equipment.

When the selected transition and transport steps in an operation are highly complex and expensive, it may be valuable to include them in the Process at a Glance, as previously described. Fully understanding the complexity, the costs, and the impact of normal material variation on the transition steps will help ensure a fast climb to standard quality and costs. Do not underestimate the challenges these subsystems may create in the final operation. The hand-offs between the value-adding processes has been the Achilles' heel in many operations.

When new technologies are incorporated, gaining run-time experience with them is critical to understanding their benefits and issues. Most engineers tend to be linear in their thinking and this can lead to spending too much time developing the wrong solution or neglecting to evaluate alternatives. They take an approach and run with it, presuming that it will work. Fortunately it often does, but getting there can take significantly more time and effort than anticipated and can jeopardize the project schedule. Application of design of experiments (DOE) can create a number of test conditions to evaluate that will help optimize equipment performance with a clear understanding of which variables have the greatest impact on performance. Embedded in every aspect of 3P is the philosophy of failing fast and failing cheap. Simply put, the sooner we can find the problem, the sooner we can understand the cause of it and develop countermeasures—and the lower the cost and timing impact will be to the project.

Potholes and Stumbling Blocks

There are a number of potential problems associated with getting started in Lean 3P design. Everyone who embarks on this journey will discover their own problems and challenges, and like learning to walk as a child, sometimes we just have to keep falling and getting up to develop the full understanding and mastery of the process. Even the word *mastery* implies that there is an endpoint where we "arrive." With Lean 3P design, as with the other elements of Lean, the simplicity of the concepts masks the richness and complexity of them, and the learning process never stops. At best you figure out how to make new mistakes, solve them more quickly, and avoid some of the ones that you made before. With that as a backdrop, here are a few of the "Eureka" moments that I have seen in applying the 3P process.

Participant Selection

Optimists do better than pessimists. The 3P process is so radically different from traditional design processes that it goes much easier when people have an open mind and a willingness to try it. Pessimists can usually be encouraged to participate and engage and sometimes even become strong proponents of the process, but do not make their conversion an unstated goal of the event. As the organization becomes more accepting of the process, there will be time to bring the naysayers along, but in spite of their potential expertise, they will detract more than benefit the process. If their knowledge and expertise is needed, consider having them participate as outside resources, available for consultation during the event.

Judges' (Critical Evaluators) Selection

Lean 3P design should challenge the status quo if it is done well. The preliminary product design may be greatly affected as the process is concurrently developed. Traditional ways of handling functions or processing steps may be displaced by new methods. The Judges should be carefully selected to include people who have the strategic business and technical knowledge to guide the 3P team participants, and they are often the key decision makers. When considering whom to include as a Judge, look at who the people are in the organization who must support the new direction and approve it. Always include these key decision makers as Judges, especially if they tend to be conservative and risk adverse. Any negativity they have will still influence the team, but it should be nominal because the Judges are only brought in at specific times to aid in the evaluation of the design concepts. The participants will already be thinking about these key decision makers and anticipating their reaction to possible changes. Having them participate as Judges allows them to experience the same journey of discovery as the participants and be influenced by the quality of the data and the enthusiasm for the new ideas. In the rare event where a good idea is "shot down" by a Judge, it still frees the team to move on to the next one with an understanding of why the decision maker chose not to endorse the original. 3P, however, is a very fact-based, data-driven process, and it is not typical to see ideas arbitrarily rejected without good reason.

Judges' (Critical Evaluators) Role

The role of the Judges can be confusing due to the title. Their function in the event is not to have final approval of the ideas presented by the team, but to listen to the concepts developed by the team and to add perspective. Occasionally they may need to be convinced to support an idea that has potential for great reward, though coupled with higher risk. Reminding them that the 3P process embodies the Plan, Do, Check, Act cycle and assures that every idea will be validated with more data before allowing it to clear the next 3P phase-gate can help gain the permission needed to go forward with further development. In every evaluation that takes place during the 3P event, the participants get to vote along with the Judges. The objective generally is not to reach complete consensus but to garner enough support for the better ideas to move forward.

Participant Pairing

It was mentioned earlier, but is worth noting again, that pairing the participants carefully during the event makes a big difference to the outcome. Sometimes personalities are a factor, and skill sets and functional knowledge are always a determinant. People who just do not get along well probably should not be paired up if it can be avoided. People who are overly dominant should not be paired up with people who tend to be subservient.

In the course of the 3P event the participants will intentionally be paired with one team, then split up and paired with other people a number of times. This is done by design to minimize their attachment to specific ideas before considering others. It forces people to work on developing the benefits of new concepts and avoids latching onto the first idea that they work on. Successful participant pairing will contribute significantly to overall success.

Let It Breathe

There is an ebb and flow to a 3P event that must be allowed to happen in spite of the time pressures imposed on the event. Time is a huge factor, and the event leaders and facilitators will need to keep a close watch on progress throughout. However, there will be times when the teams must dwell on the new concepts and consider the possibilities, speaking to subject matter experts, vendors, and doing research to confirm or dismiss ideas. Normally the biggest risk to this process is during the early development of the three prototypes. If the time is not adequate to vet the new ideas, the teams will very quickly snap back to the tried-and-true methods, parking the breakthrough idea on a bucket list never to be seen again, unless by the remote chance the selected method fails.

Time Management

The role of the facilitators and event leader must focus on time management and continually help the teams prioritize the most important items to work on. This is most difficult during the latter part of the event, when constructing prototypes and conducting complex evaluations. Teams are spread out and small pockets of people are drilling down on specific assignments, researching, looking for construction materials, and communicating with vendors and suppliers. It is very hard to keep on top of all the activities. Setting expectations for long days, short breaks, and working lunches all help maximize the amount of time during the event. Helping teams work on the right priorities and gently cutting long-winded speakers off is more difficult but is enormously important.

The event leaders and facilitators must conduct wrap-up review sessions every night after the participants leave to discuss any behavioral problems associated with the pairings of people and to reprioritize, making required adjustments to keep the event on track.

Leadership

The ability of senior management to lead benefits any organization, any time. In 3P it becomes even more necessary due to the radical nature of the process. Asking grown-ups with extraordinary skills, training, and advanced degrees to think like 12-year-olds requires a great degree of self-confidence and leadership. Having people create mock-up prototypes out of cardboard and sticks will seem ridiculous. Leaders need to provide the assurance that it will deliver the expected results in an unwavering manner.

Every 3P event will have great highs and some lows. Even after experiencing many 3P events, there always seems to be a time in every event where the outcome looks hopeless. It might be due to the way the participants are gelling together, or something one of the Judge's said that derailed a potentially great idea. The teams may be despondent and confused and helping them see the potential around the next corner will require leadership attention and action.

Set Big Expectations

Most people will only rise to the level we permit them to. In Safety, the saying goes that people will only perform to the level of your lowest expectation. Proponents of Lean 3P design all agree that you must set big goals. My belief is that the goals need to be based on the product and operation being developed through the process, and set accordingly. However, in every case, they should be stretch goals that are very challenging, but attainable, versus nominal goals that can be accomplished in traditional ways. This is the only way that the teams will have the impetus and freedom to reach that next level. Given the opportunity to try, and our faith in them to succeed, people are capable of almost anything.

Believe in the 3P Process and Stick with It

Throughout the event and especially in the weeks and months following there will be a great tendency to revert back to established approaches to developing the designs. Establishing the teams, using visual indicators, and progressing the Process at a Glance parameters will help the team keep on track; especially with the initial projects, it will be foreign and require a lot of encouragement to keep it moving forward until it becomes a standard way of working. This may be especially hard for the peripheral support groups who do not yet understand the full value of the process. As the benefits become realized and as small wins become great successes, the culture of the organization will become accepting and embrace the 3P approach.

Postevent Expectations

Before the event ends and the final report-out is given, develop an understanding with the various functions participating about the new way of working on the project. Once 3P becomes established as a viable process, this commitment may be less necessary, but until it is embedded into your culture, the idea of having the various functional groups involved in "engineering" is not generally understood or supported. Everyone in every function is busy today, so allowing people to participate in a project like this on an ongoing basis will be foreign and will of necessity require other tasks to be eliminated and dropped. Establishing that expectation up front will avoid confusion and conflict. Failure to gain this support and participation will not necessarily nullify the benefits of the 3P process, but it will take much longer to realize them. The ability to lock-step knowledge and understanding of problems and benefits among all of the functional groups will ensure great ownership of the final operation and the smoothest possible launch. It sets the stage for people to collaborate and solve problems together, and the relational behavior becomes that of collaborating to help versus setting expectations and demands typical of the traditional approach.

With all of these steps in the process complete, the 3P event is concluded, but the process is really just beginning. Over the weeks and months ahead, the teams formed during the 3P event will continue to collaborate and work through the inevitable issues that occur in all projects. There is no substitute for continuing forward with the teams and making the progress on the product and operation visual to the organization. 3P represents a true culture change and a new way of working and innovating, with extraordinary benefits that are only limited by our collective imaginations and willingness to try.

Chapter 14

Evergreen Lean 3P Design: Conclusions

Evergreen Lean 3P Design

Applying the A3 methodology and following the Lean 3P design process steps for routine design upgrades, to either product or process, may seem excessive, and for very minor changes they might be. However, introducing this approach for making change, as a routine process, should create enormous value for any organization. Clearly the benefits will be significantly greater than having individuals working independently on projects and handing them "over the fence" when completed.

It is also a powerful development tool for the entire organization, helping people from every functional area see how things work and ensuring that their functional perspective is incorporated in the future state. It is especially beneficial for developing inexperienced engineers and imparting to them Lean design concepts and the value of collaboration and try-storming to visualize ideas. The 3P concept is still very new to most companies, and applying it in an evergreen manner to small design changes is still being developed. As more companies begin to apply 3P, more techniques will surely be developed as others test and improve the efficiency of the process. For now, I hope this becomes a starting point that you will try and benefit from.

Within the Lean Community there is growing interest in applying 3P to more modest process or product design changes and solving day-to-day problems associated with the product or process that manufactures it. Practitioners of 3P all agree on the value of the process and the ability to create huge gains compared to traditional approaches. More and more people are starting to consider how it might apply to more modest improvements on an ongoing basis. It could be described as more like an "Evergreen" 3P versus an event-driven, large project based 3P process.

But what elements of 3P could work in this type of application? What aspects will not? Is the process simply the process? How scalable is 3P?

Taking a look at the key attributes of 3P we see that the balance between encouraging individual brilliance alongside group collaboration is a key factor. Another major aspect is the ability to gather the input from many different functional areas, all stakeholders in the product or process. Having people think "out of the box," brainstorming, and try-storming all seem to work on big or small projects. Thinking through how various functions occur in nature, as you did when you were younger, and then migrating toward more industrial applications of those natural principles worked for individual process steps. Why wouldn't it work for a similar smaller product or process improvement?

If the minor design project is not destined to go through a large-scale 3P event, there may not be a formal Event Charter, though all of the elements should still apply. Understanding the Goals of the proposed change is mandatory every time, though admittedly it is not always practiced well. Too frequently we jump from problem to solution without stopping to understand what really happened or what other alternatives might be available.

Clear Boundary Conditions should also be defined any time a project is started. Every design project must have set targets for costs, delivery timing, and ability to meet customer takt time and product quality. These and any other goals should be tracked visibly throughout even the most modest design change project.

Identifying up front who the major Participants should be in contributing to a problem resolution seems to make sense as well. These are the stakeholders who are being impacted by a given problem now, and they have a vested interest in improving the conditions and outcome, even if they personally do not have the skills necessary to impact it directly. This is another way that the 3P process opens doors for improvement that allow everyone to contribute to the solution. Outside resources, though one step removed from the actual 3P event, will also be involved in almost every project.

The Judges fill a big role in the large 3P event. They provide strategic insight and because many times they will also be key decision makers, they are enablers who give the design team permission to take calculated risk with a concept that will likely provide big benefits. Without their endorsement, many great ideas would never be pursued out of fear that they would either be shot down, or worse, that the engineer would fail attempting it, and suffer the consequences—real or perceived. This is one of the least appreciated benefits of 3P. Risk is defined, evaluated against potential benefits, and supported by all of the stakeholders and key decision makers. This facet of 3P is a great enabler that opens up the potential for major breakthroughs. With smaller design change projects the make-up of the Judges may be different from very large ones, but the nature of the support and the need for input into the intended change are critical for any design project.

The concept of Process at a Glance captures all of the key aspects in the design of virtually any type of equipment. Defining the Material being processed, along with the normal variation, and visually depicting the Method of how the material will be processed in a value-adding step is a routine process. Too often, however, it is not documented in a way that is both visual and readily understood beyond the individual developing it. Considering how to Gauge and verify that the functional step has been performed correctly and that the new state is defect free makes sense in any application. Some processes will require a Tool or a mold to accomplish the functional step, and many will also need some type of Fixture or Jig to hold a part or parts in place while the operation is being performed. Lastly, the Machine considerations for guarding, personnel protection, microbiological counts, cross-contamination concerns, and similar issues should apply in some manner to any change. This suggests that Process at a Glance applies very well to both large and small design changes.

Developing Evaluation Criteria for 3P from the 30 suggested options forces consideration of what is *really* important in the final design. It is pretty clear that many of them work very well with large-scale operational changes or when developing completely new operations, but they might not be well-applied to smaller design changes. However, even in a major 3P event, the goal is not to evaluate based on all of the criteria, but to limit them to a more manageable number that is significant to the task at hand. With that backdrop, going through all 30 and thinking through how they might apply, and which ones are most meaningful to the business for the change being considered, seems like a process that could only add value. In fact this relatively quick analysis should benefit all projects, big or small.

Developing physical Prototypes fast and at low cost becomes much simpler for smaller changes, yet most engineering teams tend to start with a modification of a CAD drawing to develop concept. Any time a physical model can be developed it takes the understanding of the problem and multiplies it compared to looking at a 3D CAD layout. In fairness, 3D CAD can often be used in conjunction with the physical prototype, but spending a few hours to build a mock-up of the intended design in order to communicate the idea will be more effective, quicker and less costly than using CAD. Changes to the physical prototype can be made by simply creating a foam part, moving a catch bin, or sawing a few inches off of a stick leg. And the whole team gets to touch, see, and participate in every change.

The 3P report-out session at the end of an event gives the summary of all the key lessons learned and defines the next action steps for delivering the project. It then receives the approval from the Judges and support of the Participants to go forward with the plan. Traditional engineering has Concept Reviews and Design Reviews that attempt to serve that same purpose. Small projects still need a form of this consensus and approval step in order to be progressed. The full 3P event report-out may not be needed for a minor project; however, the principles and value derived from it always apply.

Reviewing all of the key elements of 3P, it appears like all of them apply, to some degree, to any design project. But how do you put it all together into a comprehensive process that flows? For small product or process changes a multiday event with a large team does not work. The Evergreen 3P concept is about taking the core benefits of 3P and applying them to the way routine work is performed with great collaboration, functional representation, rapid learning, alternatives, and prototyping, all to hone in on integrated solutions that benefit product, process, and people.

The A3 Process Applied to Small-Scale 3P Design

As discussed earlier A3 is a means of solving problems by telling the story succinctly on a single sheet of paper, encompassing the problem statement, background, current condition, goals, analysis, countermeasures and action plans, with some plan for follow-up. A3 embodies the Plan, Do, Check, Act continuous improvement cycle. It is the perfect vehicle for coaching at every point in the development of the document and is intended to be a collaborative methodology. The A3 tells the story and is typically supported with additional detailed documentation referenced in the A3 report.

Applying the A3 process to smaller design changes makes sense with or without the application of 3P design techniques. Using it in conjunction with

3P provides a structured approach that helps a team derive the benefits and provide clarity to the process.

Problem Statement

A statement that succinctly explains the deviation from the norm needs to be corrected. This should be a starting point for any redesign of product or products.

Background

The background provides perspective on how big or small the problem is and provides context.

Additional considerations from a 3P perspective: Define the stakeholders and key decision makers who are affected (participants). Consider who or what is affected and quantify the significance of the issue.

Current Condition

States the immediate impact of the problem and any associated activities.

Additional considerations from a 3P perspective: Consider what boundary conditions are there that must be adhered to. Determine outside resources that could be available to help understand the problem and help develop solutions.

Goals

Goals are the new desired conditions that the countermeasures must deliver to be successful. Boundaries (time, money, resources, space, etc.) are stated.

Additional considerations from a 3P perspective are as follows: Consider required customer demand takt time. Consider the list of 30 potential 3P evaluation criteria and agree ahead of time on the ones that are most important.

Analysis

Determine the root cause and contributory causes of the undesirable condition.

Additional considerations from a 3P perspective are to go to the workplace where the problem is occurring *with* the stakeholders and look at the actual current condition. Discuss the issue or issues and collect data on when the undesired conditions occur, how often, and if there are any other patterns to support the analysis. Avoid asking who, but instead focus on the *process* that failed. If a process is robust the people aspect of it should also be robust. Recognize that "human error" is a by-product of a poorly designed process.

Countermeasures

Countermeasures mitigate the causes of the problem and enable you to move toward the goal state.

Additional considerations from a 3P perspective are as follows: This section of the A3 is where many of the elements of 3P get applied:

- Using a team approach, involve the participants by identifying the product attribute or process value-added function that needs improvement. If applicable, lay out the parts and materials to gain understanding of how they work together. For product redesign, understand and rate the features in question. Develop prototypes for evaluation.
- With the value-added process function defined consider seven alternatives in nature.
- Migrate the alternatives from natural phenomena to industrial applications to the extent possible. Consider other industrial mechanisms. Determine if there are exemplars in your own business that should also be considered.
- Use the evaluation criteria developed in the Goal section to evaluate the alternatives using the Pugh method to compare them.
- Select the three best Alternatives and develop the simple Process at a Glance for each one. Note that for minor redesign projects you may only have one or two columns representing the value-adding steps.
- Have the participants go and create three low-cost prototypes, completing the sections in the Process at a Glance that are possible with the information you develop. It might be necessary to call in vendors or other subject matter experts to augment the team in developing the application of new technologies. If the team is large enough, break up into teams to divide the work.
- With three alternative prototypes developed and three different Process at a Glance charts, use the evaluation criteria and with participants and Judges rate them and select the best features from each.

- Form one team again and develop the final prototype from the three models, using the best aspects of each. Build great amounts of detail into the final prototype. It takes little time and everyone can add value.
- Update the Process at a Glance with the final value-added functions. Now include the transition or transport steps into the Process at a Glance chart, evaluating them for complexity, takt time, error-proofing, and any quality checks that might be needed to verify that the value-added step was done correctly.

Action Items

Clearly state what will be done, who is responsible, and when it will be complete.

Additional considerations from a 3P perspective are to employ a stand-up meeting in front of the prototype model to review actions and report-out information.

Follow-Up

State the plan for measuring the effect of the countermeasures in achievement of the goal.

Additional considerations from a 3P perspective are that the time required to reach the standard quality rates and standard costs should be very short, hours to days, versus weeks to months using traditional approaches.

Conclusions

Lean 3P design is possibly the simplest and most powerful approach to structured innovation and design ever developed. It combines individual innovation with cross-functional collaboration. Lean 3P is a key component of an overall Lean Product Development process. It includes all relevant stakeholders of a product or process and gets input at the earliest possible moment in a way that allows the maximum flexibility in the shortest time, with the lowest cost. 3P achieves dramatic, revolutionary results and assures that a vast number of ideas are generated and quickly progressed. It further drives to ensure that only the very best concepts get through the evaluation gates to earn the right to be developed. It is exciting, energetic, and fun, and incorporates the ideas of all participants irrespective of rank and position in the organization.

Although this text is intended to be a "how to" guide to learning and applying 3P to your product development for new operations, it is important to emphasize that 3P is not in itself a standard process. Rather, it is an extremely pliable system for developing ideas quickly and efficiently. Some aspects of 3P will work perfectly with your products and operations and others may not. This is a 3P process guide, not a cookbook.

Comparing a novice cook, who accurately measures every ingredient in a recipe, to the renowned chef who randomly grabs and tosses handfuls of ingredients around while masterfully creating delicacies, early practitioners of 3P will want to follow these steps as closely as possible, learning and adjusting the process for their business as they grow in experience. 3P is a variation of the scientific method, an extension of Plan, Do, Check, Act, the continuous cycle of improvement that refines every good process into a better one. The first few times around the circle will feel uncomfortable and not without a few mistakes.

The most critical step in the process is the first decision to just try it. As noted before, a good understanding of the Lean toolkit and Lean thinking by most of the participants is an important prerequisite but should not be a deterrent, simply a necessary first and highly valuable step. Many people are galvanized by the concept of potential gains to be had from 3P, but then find excuses for why it might work for "them," but will not work for "us." Like a lot of the simple tools taught in Lean workshops, 3P is relatively easy to understand, but underneath the surface of the simple concepts are profound truths that when embraced, deliver enormous value.

Lean 3P has the power to change the way we work and develop projects in a manner that eliminates the need to retrospectively use many of the Lean tools, normally applied to improve established processes. Total Productive Maintenance and Quick Changeover are built in along with Flow, Kanban, Jidoka, and Poka Yoke. The housekeeping and organizational elements of 5S are baked into the new operation along with other criteria deemed important to your plant or business. 5S more strategically is a tool to expose abnormality that will continue to render benefits long after product launch. This is not to say that other continuous improvement stops after an operation is designed using 3P, but it is probable that future improvements will be less dramatic than the ones initially found when examining established operations during typical Kaizen events.

The ability to use 3P in any industry or business, with any product type and at any scale, is certainly implied if not a given. The scale of the prototypes and length or number of the 3P events might change, but the concepts seem to apply anywhere, and more and more companies are validating that premise every year. Like many of the techniques used in Lean, 3P can be applied with different levels of understanding and commitment. The immediate benefits resulting from conducting a 3P event for a new product and operation will be seen over the life of that product. Pursuing 3P at much deeper levels will not only drive continuous improvement in Operations and create value for the business, but it will be a powerful process that encourages individual creativity and innovation, organizational collaboration, calculated risk-taking to achieve breakthrough results, and a means of affecting positive change that embraces all of the tools in the Lean toolkit.

In writing this it is my sincere hope that the reader will gain a better understanding of the potential of this amazing methodology and have the courage to try it. There is no doubt that it will provide a competitive advantage to those who do, with the more strategic advantage going to those who embed 3P deeply into their culture as new products and operations are designed. The 3P process effectively breaks down departmental silos and exposes new ways of improving battered inefficient processes. It signals to the organization that invention and innovation can happen anywhere, even on the shop floor, and provides the structure to embrace creativity with the necessary safeguards to mitigate risk and ensure that change controls are in place and utilized.

As more and more companies embrace the concepts of 3P and Lean, we will find ways of making sustainable improvements to many of the problems facing us. I wish you every success along your journey.

Appendix

Lean 3P Event Charter Template

Process Name	Start Date	End Date
Process Boundaries	Start Time	End Time
	Event Meeting Location	
	Process Owners	
Why Do We Need This Event?	Process Experts	
Event Targets	Event Judges	
	Process Outsiders	
	Facilitators	

Flow Diagram of Lean 3P Design Process

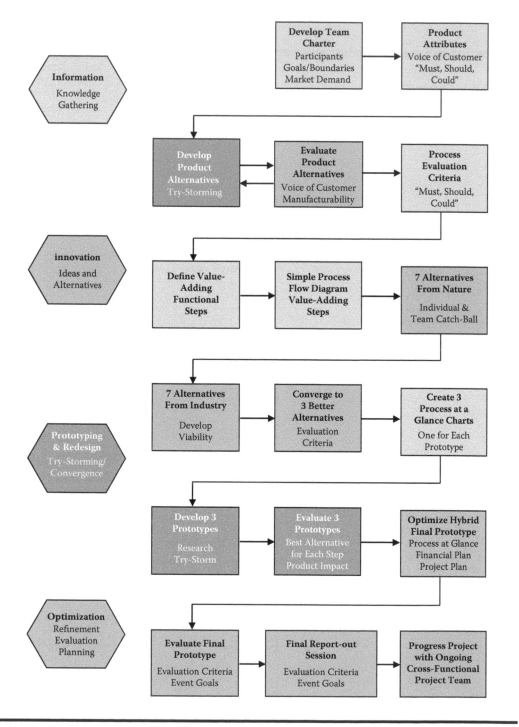

Figure AP.1 Flow diagram.

Lean 3P Event Materials Checklist

Board stock (good for adding structure or forming brackets)		
Yes?	*Amount*	*Material*
☐	_____	2" × 4"s
☐	_____	1" × 2"s
☐	_____	1" × 3"s
☐	_____	Sheet stock
☐	_____	4" × 8" sheets of Luan or Chipboard
☐	_____	Cardboard—almost any size (4" × 4" "slip-sheets")
Fasteners		
Yes?	*Amount*	*Material*
☐	_____	Nails (different sizes) _____
☐	_____	Drywall screws (different sizes) _____
☐	_____	Staples
☐	_____	Duct tape
☐	_____	Painter's tape
Miscellaneous		
Yes?	*Amount*	*Material*
☐	_____	Markers
☐	_____	3/4" or up to 1½" PVC pipe
☐	_____	"Boneyard" materials _____
☐	_____	Old computer monitors
☐	_____	Canvas
☐	_____	Pieces of wire
☐	_____	Tables
☐	_____	Other? _____

Tools and Safety		
Yes?	*Amount*	*Material*
☐	_____	Safety glasses
☐	_____	Gloves
☐	_____	Safety shoes
☐	_____	Chop saws
☐	_____	Handsaws
☐	_____	Hacksaws
☐	_____	Circular saw
☐	_____	Razor knives
☐	_____	Staple guns
☐	_____	Hammers
☐	_____	Drills
☐	_____	Hole bits
☐	_____	Extension cords
☐	_____	Screw guns

Lean 3P Pre-Event Activity Checklist

Two to Three Months Before
☐   Arrange Facilitators
☐   Invite Judges (Critical Evaluators)
☐   Develop cross-functional participant list
☐   Invite participants
☐   Consider government grant request
☐   Plan Lean refresher training
☐   Plan Lean 3P orientation and schedule
One to Two Months Before
☐   Finalize Event Charter with Process Owner's approval
☐   Review Event Charter with Judges and Key Participants
☐   Reserve event space
☐   Develop list of materials and tools
☐   Obtain children's books for reference
☐   Information gathering—marketing data, Voice of the Customer, etc.
☐   Conduct Lean refresher training
☐   Conduct 3P orientation meeting with participants
One to Two Weeks Before
☐   Confirm attendee participation
☐   Arrange meals and snacks for event
☐   Purchase and arrange delivery of materials and tools
☐   Prepare event space—identify physical boundaries
☐   Develop event "contract"
☐   Divide participants into three teams

The Day Before	
☐	Review set-up of event space
☐	Verify materials and tools
☐	Print copies of contracts
☐	Verify video projector, flip charts, markers, plotter paper, whiteboards
☐	Collect product and material samples, packaging options, literature, product descriptions, specifications, etc.

Potential Design Evaluation Criteria

Evaluation Criteria	Definition
Takt time	The pace of an operation, producing no more and no less than customer demand requires. Takt enables synchronized production, with each step scaled to make and feed materials from one station to the next in the same time, ideally without inventory buffers.
One-piece flow	Moving one part or workpiece at a time through the production process in a system that flows from start to finish.
Pull system	An upstream process creates a signal for a downstream process to produce the next part or lot of material. Pull reduces the need for inventory buffers between process steps.
People involvement	Encourages higher levels of people interaction and engagement in the process, seeking to optimize the balance between people contributions in the performance of standard work, versus automation both in terms of costs (initial and ongoing) and flexibility.
Automatic unloading	A feature built into the operation that automatically processes the last part through to always leave the machine available for the next run. It can reduce confusion and errors and decrease setup time.
Load-load operations	Operators physically travel, taking the part from one machine and loading it into the next. Load-load engages the operator, often providing a secondary function, such as a visual quality check.
Low-cost automation	Focus on achieving the benefits of automation at a fraction of the typical costs. Utilize gravity conveyors versus powered conveyors; avoid overbuilding of brackets or structural elements, when simpler designs are functionally adequate.
Mistake-proof (Poka Yoke)	Make it nearly impossible to produce poor-quality product by designing components that prevent it. Physical or mechanical solutions that only enable positioning in the correct orientation, sensor technologies or bar codes can all be used to mistake-proof.
Minimal capital	Use creativity to develop solutions that are highly functional at a low capital cost. Consider increasing people involvement versus more automated solutions. Strong focus on innovative solutions.

Minimal space required	Design based on compactness to minimize travel distances and motion waste as well as costs.
Low-motion waste	Low motion to access tools, materials, control panels, and inspection points. Optimize both the movement of product and the people who are operating the process.
100% Gauging	Add checks at each processing step to ensure that quality problems are detected immediately. This metric utilizes measurement tools, sensors, and other means of instantly checking and verifying quality.
Maximum operator value-add	Engage operators in the value-adding processes. Reduce travel time retrieving parts or materials, conducting routine inspections, writing up reports, or other lower level work.
Changeover time	Reduce frequency and lost time from changeover by designing in quick disconnects, pinning, making required tools accessible and organized. Utilize pre- and postchangeover times to perform associated work when possible.
Tool room maintenance	Develop systems to manage key tools such as molds so that their status is visible, with color-coding systems or physical locations for tools that are often required to improve productivity.
Tool quality or tooling cost	Design or purchase tools that are fit for the required usage. Consider high quality tools if appropriate or inexpensive disposable tools when quality isn't required, or if tool cleaning costs are excessive.
Safety, ergonomics, and health	Accommodate requirements for personnel safety and meeting all regulatory requirements or corporate initiatives. Consider protection of employees above other design factors.
Environmental impact	Accommodate all regulatory requirements or corporate initiatives. Consider environmentally friendly initiatives, recycling, and waste minimization.
Internal waste collection	Design to effectively remove waste streams automatically from the process. Segregate waste close to the point of generation to gain benefits, versus a downstream, commingled approach.
Simple as possible	Develop single-purpose operations where more value is placed on ensuring the function be done correctly every time, versus the benefits gained by flexibility or perhaps better flow.

Standard or off-the-shelf equipment	Application of standard equipment to increase familiarity, reduce documentation time and enhance maintenance repair and the availability of spare parts. Training materials are often available and proven.
Process capability (Cp)	Design for minimal process variation and a consistent process to maintain high quality and throughput rates.
Knows process	Incorporate part or all of an existing process into the design of the new operation if there is technical or strategic benefit. Focus on optimizing other aspects of operation.
Future challenge	Consider potential regulatory or intercompany change that may happen. Anticipate changes and incorporate into the design to benefit with lower compliance costs, ease of relocation, or adoption of a new technology.
Maintenance free	Design in the ability to quickly service equipment and make it robust enough to greatly extend time between failures. Consider on-line redundancy, lubrication points, access panels, and choice of materials of construction to ensure long equipment life and minimal down time.
Technical advantage	Create a barrier for entry by a competitor using a different technology that could leap-frog your company due to lower costs, better ease of use, or more consistent, higher-quality product.
Autonomation	An automated machine's ability to detect a quality defect, stop itself, and subsequently alarm and notify an operator to respond and address the problem on a timely basis.
Development time or in-house development	Design in a way that reduces project time. Consider application of known technologies, reduced automation, or the way computer-aided design (CAD) is done. In-house design versus external can reduce costs and increase consistency but is more resource-intense.
Scalability	Enable the process to operate efficiently at a fraction of the full capacity through to multiples of the target rate. Avoid overbuilding at a higher cost.
Flexibility	Design with great flexibility to share or store equipment, allowing the manufacturing space to be utilized more effectively or ensuring higher utilization of people or equipment.

Potential Design Evaluation Criteria Checklist

	Evaluation Criteria	*Must*	*Should*	*Could*
1	Takt time			
2	One-piece flow			
3	Pull system			
4	People involvement			
5	Automatic unloading			
6	Load-load operations			
7	Low-cost automation			
8	Mistake-proof (Poka Yoke)			
9	Minimal capital			
10	Minimal space required			
11	Low motion waste			
12	100% gauging			
13	Maximum operator value-add			
14	Changeover time			
15	Tool room maintenance			
16	Tooling quality or tooling cost			
17	Safety, ergonomics, and health			
18	Environmental impact			
19	Internal waste collection			
20	Simple as possible			
21	Standard or off-the-shelf equipment			
22	Process capability (Cp)			
23	Known process			
24	Future challenge			
25	Maintenance free			
26	Technical advantage			
27	Autonomation			
28	Development time or in-house development			
29	Scalability			
30	Flexibility			

Note: Evaluate each of the criteria and check the appropriate box: *Must* Use, *Should* Use, *Could* Use.

Seven Alternatives with Evaluation Criteria Template

Seven Process Options for:								
Material								
Method								
Gauge								
Tool								
Fixture or Jig								
Machine								
Evaluation Criteria	*Weighting*							
1								
2								
3								
4								
5								
6								
7								
8								
9								
10								
Total +								
Total –								
Rating (1 through 7)								

Seven Alternatives Template

Seven Process Options for:							
Material							
Method							
Gauge							
Tool							
Fixture or Jig							
Machine							

Process at a Glance Template

Process at a Glance										
Process Step	*1*	*2*	*3*	*4*	*5*	*6*	*7*	*8*	*9*	*Complete Process*
Material										
Method										
Gauge										
Tool										
Fixture or Jig										
Machine										

Process at a Glance with Evaluation Criteria Template

												Complete
Process at a Glance												
Process Step		1	2	3	4	5	6	7	8	9		*Complete Process*
Material												
Method												
Gauge												
Tool												
Fixture or Jig												
Machine												
Evaluation Criteria	*Weight Factor*											
1. Minimal capital												
2. Changeover time												
3. Low-motion waste												
4. Pull system												
5. Autonomation												
6. Off-the-shelf equipment												
7. High process capability												
8. Internal waste collection												
9. Simple as possible												
10. Maintenance free												
11. Scalability												
Total +												
Total −												
Rating total												

Works Cited and Bibliography

Ackoff, Russell Lincoln, Jason Magidson, and Herbert J. Addison. *Idealized Design Creating an Organization's Future*. Upper Saddle River, NJ: Wharton School, 2006.

Altshuller, Genrich. *Innovation Algorithm*. Worcester, MA: Technical Innovation Canter, 1973.

Altshuller, Genrich. *Creativity as an Exact Science*. New York: Gordon & Breach, 1984.

Altshuller, Genrich. *And Suddenly the Inventor Appeared* (translated by Lev Shulyak). Worcester, MA: Technical Innovation Center, 1994.

Deming, William Edwards. *Out of the Crisis*. Cambridge, MA: Massachusetts Institute of Technology, Center for Advanced Engineering Study, 1982.

Dennis, Pascal, and Jim Womack. *Getting the Right Things Done: A Leader's Guide to Planning and Execution*. Cambridge, MA: Lean Enterprise Institute, 2006.

Galsworth, Gwendolyn D. *Visual Workplace: Visual Thinking*. Portland, OR: Visual-Lean Enterprise, 2005.

Galsworth, Gwendolyn D. *Work That Makes Sense: Operator-Led Visuality*. Portland, OR: Visual-Lean Enterprise, 2010.

Gelb, Michael, and Sarah Miller Caldicott. *Innovate Like Edison: The Success System of America's Greatest Inventor*. New York: Dutton, 2007.

Goldratt, Eliyahu M. *Critical Chain*. Great Barrington, MA: North River, 1997.

Hino, Satoshi. *Inside the Mind of Toyota: Management Principles for Enduring Growth*. New York: Productivity, 2006.

Keyte, Beau, and Drew Locher. *The Complete Lean Enterprise: Value Stream Mapping for Administrative and Office Processes*. New York: Productivity, 2004.

Koenigsaecker, George. *Leading the Lean Enterprise Transformation*. Boca Raton, FL: CRC, 2009.

Liker, Jeffrey K., and David Meier. *The Toyota Way Fieldbook: A Practical Guide for Implementing Toyota's 4Ps*. New York: McGraw-Hill, 2006.

Locher, Drew A. *Value Stream Mapping for Lean Development: A How-To Guide for Streamlining Time to Market*. New York: CRC, 2008.

Locher, Drew. *Lean Office and Service Simplified: The Definitive How-To Guide*. Boca Raton, FL: CRC/Taylor & Francis Group, 2011.

Mascitelli, Ronald. *The Lean Design Guidebook: Everything Your Product Development Team Needs to Slash Manufacturing Cost*. Northridge, CA: Technology Perspectives, 2004.

Mascitelli, Ronald. *The Lean Product Development Guidebook: Everything Your Design Team Needs to Improve Efficiency and Slash Time-to-Market*. Northridge, CA: Technology Perspectives, 2007.

Mascitelli, Ronald. *Mastering Lean Product Development: A Practical, Event-Driven Process for Maximizing Speed, Profits and Quality*. Northridge, CA: Technology Perspectives, 2011.

Pugh, Stuart. *Total Design Integrated Methods for Successful Product Engineering*. Wokingham, UK: Addison-Wesley, 1990.

Rolfes, K., KDR Associates presentation, AME Conference, Australia, August 15, 2007.

Shingō, Shigeo. *A Revolution in Manufacturing: The SMED System*. Stamford, CT: Productivity, 1985.

Shook, John. *Managing to Learn: Using the A3 Management Process to Solve Problems, Gain Agreement, Mentor and Lead*. Cambridge, MA: Lean Enterprise Institute, 2009.

Sobek II, Durward K., and Art Smalley. *Understanding A3 Thinking: A Critical Component of Toyota's PDCA Management System*. Boca Raton, FL: CRC, 2008.

Ward, Allen C. *Lean Product and Process Development*, Cambridge, MA: Lean Enterprise Institute, 2009.

Womack, Jim. *Gemba Walks*, Cambridge, MA: Lean Enterprise Institute, 2011.

Index

The Author

Allan R. Coletta is a chemical engineer with an extensive background in manufacturing operations, supply chain, and engineering, gained while working in the chemical process and health-care diagnostics industries. His Lean experience started while serving as site manager for ICI Uniqema's largest specialty chemicals plant in North America (New Castle, Delaware) and continued to expand his role as Senior Director of Engineering for Siemens Healthcare Diagnostics (Newark, Delaware). His passion for manufacturing and engaging people in continuous improvement continues to grow through personal application of Lean principles.

Allan serves on the Delaware Manufacturing Extension Partnership's Fiduciary and Advisory Boards and is a member of the Delaware Business Mentoring Alliance. He is also a member of the American Institute of Chemical Engineers and the Association for Manufacturing Excellence (AME).